NATURAL RESOURCE MANAGEMENT

Volume 8

After the Green Revolution
Sustainable agriculture for development

Full list of titles in the set
Natural Resource Management

After the Green Revolution
Sustainable agriculture for development

Gordon R. Conway and Edward B. Barbier

Routledge
Taylor & Francis Group

LONDON AND NEW YORK

First published in 1990

This edition first published in 2009 by Earthscan

ISBN 978-1-84971-019-0 (hbk Volume 8)
ISBN 978-0-415-84594-6   (pbk Volume 8)

ISBN 978-1-84971-011-4 (Natural Resource Management set)
ISBN 978-1-84407-930-8 (Earthscan Library Collection)

For a full list of publications please contact:

Earthscan
2 Park Square, Milton Park, Abingdon, Oxon OX14 4RN
Simultaneously published in the USA and Canada by Earthscan
711 Third Avenue, New York, NY 10017
Earthscan is an imprint of the Taylor & Francis Group, an informa business

Earthscan publishes in association with the International Institute for Environment and Development

A catalogue record for this book is available from the British Library

Library of Congress Cataloging-in-Publication Data has been applied for

Publisher's note
The publisher has made every effort to ensure the quality of this reprint, but points out that some imperfections in the original copies may be apparent.

At Earthscan we strive to minimize our environmental impacts and carbon footprint through reducing waste, recycling and offsetting our CO_2 emissions, including those created through publication of this book.

Gordon R. Conway was the Director of the Sustainable
Agriculture Programme of the International Institute for
Environment and Development (IIED). He is now the
representative for the Ford Foundation in charge of India,
Sri Lanka and Nepal.

Edward B. Barbier is Associate Director of IIED/UCL
London Environmental Economics Centre.

AFTER THE GREEN REVOLUTION

Sustainable Agriculture for Development

Gordon R. Conway and
Edward B. Barbier

Earthscan Publications Ltd LONDON

First published in 1990 by
Earthscan Publications Ltd
3 Endsleigh Street, London WC1H 0DD

British Library Cataloguing in Publication Data
Conway, Gordon R.
    After the green revolution: sustainable agriculture
    for development.
    1. Developing countries. Agricultural industries.
    Economic development
    I. Title II. Barbier, Edward B.
    338.1091724

ISBN 1-85383-035-6

Production by David Williams Associates (01-521 4130)
Typeset by Rapid Communications Ltd, London WC1

Earthscan Publications Ltd is a wholly owned and editorially
independent subsidary of the International Institute for
Environment and Development (IIED).

Contents

Acronyms and Abbreviations

ACORDE	Association for the Co-ordination of Development Resources
AEA	agroecosystem analysis
AKRSP	Aga Khan Rural Support Programme
BULOG	GOI procurement policy
CFC	chlorofluorocarbon
CGIAR	Consultative Group on International Agricultural Resources
CIAT	Centro Internacional de Agricultura Tropical
CIMMYT	Centro Internacional de Mejoramiento de Miaz y Trigo
EEC	European Economic Commission
FAO	Food and Agriculture Organization
FSR&E	farming systems research and extension
GDP	gross domestic product
GOI	Government of Indonesia
HYV	high yielding varieties
IARC	International Agricultural Research Centre
ICARDA	International Centre for Agricultural Research in Dry Areas
IDS	Integrated Development Systems
IFAD	International Fund for Agricultural Development
IFPRI	International Food Policy Research Institute
ILO	International Labour Organization
IMF	International Monetary Fund
IPM	Integrated Pest Management
IRRI	International Rice Research Institute
MCP	multiple cropping project
NGO	non-governmental organization
NPR	nominal protection rate

OFR	on-farm research
OFR/FSP	on-farm research with a farming systems perspective
PVO	private voluntary organization
RAZ	rapid agroecosystem zoning
SSI	semi-structured interview

Preface

This book was written while both of us were on the staff of the International Institute for Environment and Development (IIED) and grew out of a report commissioned for the United States Agency for International Development. It is based on the experiences of IIED's Sustainable Agriculture Programme between 1986 and 1989, drawing heavily on policy discussions and documents, on original research reviews, and on field work – most notably in Indonesia, Thailand, Sudan, Nepal, Pakistan, Ethiopia and Kenya.

In addition to financial support from USAID we gratefully acknowledge grants and contracts from the Swedish International Development Authority, the Ford Foundation, the World Bank, the Aga Khan Foundation, Intercooperation and the Swedish Red Cross.

We are also grateful for comments on earlier drafts supplied by Robert Blake, Robert Chambers, Johann Holmberg, Mary Lou Higgins, Jenny McCracken, David Pearce and Jules Pretty. We are indebted to Richard Sandbrook, Director of IIED in London, to Walter Arensberg of the then North American office of IIED for his support and encouragement, and to Melanie Salter for assisting the work in many ways.

Dedicated to
Thelma and Cyril Conway
and Marietta Barbier Falzgraf

Introduction

Aid to the Third World is subject to fashions. Some are trivial and short-lived; but others are longer lasting and reflect deep concerns about the nature and direction of development. Today's fashion is "sustainable agriculture". Whether it has any permanence remains to be seen, but there is no doubt that it has attracted attention throughout the aid community – development workers in the field, as well as researchers, academics, and the policy makers of the development agencies. This book is about the place of sustainable agriculture in the context of development and the steps which need to be taken if it is to be more than a passing fashion.

Inevitably, given such widespread interest, the phrase "sustainable agriculture" is open to many interpretations:

- For agriculturalists it embodies a desire to consolidate and build upon the achievements of the green revolution. They equate sustainability with food sufficiency, and sustainable agriculture can embrace any means toward that end.
- For environmentalists, though, the means are crucial. Sustainable agriculture represents a way of providing sufficient food and fibre that complements and, indeed, enhances our natural resource endowment of forests, soils and wildlife. For them, sustainability means a responsibility for the environment – a stewardship of our natural resources.
- For economists, sustainability is a facet of efficiency, not short-run efficiency alone, but the use of scarce resources in such a fashion as to benefit both present and future generations.

- Finally, sociologists see sustainable agriculture as a reflection of social values. They define it as a development path that is consonant with traditional cultures and institutions.[1]

These are very different, and in some cases contrasting, interpretations, but in the last few years the disciplines and interest groups they represent have come together to promote sustainable agriculture in a manner that has proven highly effective. Today, sustainable agriculture is widely accepted as a goal to be incorporated explicitly into policy papers and project designs. Unfortunately, though, this coalition of interests, rather than clarifying the subject, has tended to blur concepts and definitions even further. Virtually everything that is perceived as being "good" or benign is included under the umbrella of "sustainable agriculture":

- high, efficient and stable production
- low and inexpensive inputs, in particular making full use of the techniques of organic farming and indigenous traditional knowledge
- food security and self-sufficiency
- conservation of wildlife and biological diversity
- preservation of traditional values and the small family farm
- help for the poorest and disadvantaged (in particular those on marginal land, the landless, women, children and tribal minorities)
- a high level of participation in development decisions by the farmers themselves.

Many, if not all, of these goals are commonly considered to be desirable. But, as those with practical experience of development know, while it is relatively straightforward to attain one or two such goals, it becomes progressively difficult as more and more objectives are included in programme and project designs. There are trade-offs, in terms of labour, time, skills and capital, for the project and its staff, and for the farmers themselves. Choices have to be made – productivity at the expense of equity, for example, or sustainability at the expense of productivity.

Not surprisingly, attaining sustainable agriculture as currently defined is a difficult task.

After the green revolution

One cause of confusion is the fact that sustainable agriculture represents a new, and as yet barely tried, phase in development thinking. It is an important and significant departure from approaches previously associated with the green revolution; but the development community is still at that crucial transition stage where it has a notion of what the ultimate goal should be, but has yet to develop a clear, logical framework, or coherent methodological approach, for its practical implementation.

Agricultural development thinking in the 1960s and 1970s was preoccupied with the problem of feeding a rapidly increasing world population. Then, the obvious solution was to increase per capita food production. The resulting green revolution has had a dramatic impact on the Third World, particularly in terms of increasing the yields of the staple cereals – wheat, rice and maize. However, despite impressive results, it also suffers from problems of equity and failures in achieving stability and sustainability of production. For example, the new technologies are less suitable to resource-poor environments; farmers with small or marginal holdings have, on the whole, benefited less than farmers with larger holdings. Intensive monocropping has also made production more susceptible to environmental stresses and shocks. And now, there is growing evidence of diminishing returns from intensive production with high-yielding varieties (HYVs). Moreover, it has become clear that these are not simply second or third generation problems capable of being solved by further technological adjustments. They require an approach that is equally revolutionary, yet very different in its conceptual and operational style.

However, if sustainable agriculture is to be more than just a slogan of the post green-revolution era, more than a broadly stated objective open to as many interpretations as there are practitioners, then we need to define as clearly as possible what this means in both theory and practice. It is to this purpose that our book is addressed.

Sustainable agriculture for development

This book emphasizes three main themes. First, the incorporation of sustainability of agricultural production as a development objective requires explicit recognition and understanding of the trade-offs involved with other objectives. During the green revolution, maximizing agricultural yields was the paramount objective. But this was attained without sufficient attention to the sustainability and stability of production, or to how the benefits were distributed. In the post green-revolution era, all these objectives are important, yet practical experience shows that it is by no means easy to combine high sustainability with high productivity, stability and equity. Often there are severe trade-offs which, if they are to be overcome, require explicit recognition and analysis.

Second, the problems confronting the sustainability of agricultural systems are not confined to just one hierarchical level – local, national or international. Agricultural systems do not exist in isolation. They are linked across these hierarchical levels. Local production systems are tied by markets and by agro-ecological zones, to regional production areas, which in turn are linked to the national level and to the outside world through, among other things, international trade. Thus shifts in world prices or in national agricultural policies can exert powerful influences on the livelihoods of farming households. Similarly, changes in global climates, droughts and floods, pest and disease epidemics, and other large-scale calamities, have a profound impact on local production. In the opposite direction, the numerous decisions of individual households in pursuit of secure livelihoods, cumulatively affect the agricultural production of nations as a whole.

These interlinkages are not simple. The behaviour of higher levels in the hierarchy cannot be reduced to the sum of behaviours at lower levels, nor are the latter the simple disaggregate of the former. This has practical consequences; desirable interventions at one level will not necessarily have beneficial effects at another. Agricultural development cannot be based solely, or largely, on genetic engineering, or macro-economic policy, or even on farming systems research. Instead, the uniqueness of each production system in the agricultural hierarchy, and the hierarchical linkage between the different levels, mean that the problems confronting sustainable agriculture must be tackled, in a concerted fashion, at all levels – local, national and international.

Finally, putting these two themes together, proper analysis of sustainable agriculture for development requires a consideration of the trade-offs between sustainability and other development objectives among, as well as within, the different levels of the agricultural hierarchy. For example, the macro-economic goal of stimulating increased productivity of an agricultural commodity must be weighed not only against the overall national objective of sustaining agricultural development, but also against the impacts of such a policy, (and the pricing and other instruments used to implement it), on the sustainability, equity and stability of local production systems. Equally, the development of new farming systems at the local level – to overcome an environmental stress such as soil erosion, for example – must take into account overall national objectives, such as the need to earn foreign exchange, if these new systems are to be successfully adopted. In other words, the sustainability of agricultural development will depend on the analysis of each level in the agricultural hierarchy, both in its own right and in relation to the other levels above and below, with this totality of understanding then being used as the basis of development.

Outline of the book

These themes form the basis of the theory and practice of sustainable agriculture discussed in this book. The heart of the book examines the priorities and conditions for improving agricultural sustainability in developing countries:

- at the international level, focusing on the constraints of trade and the global economic order (Chapter 3)

- at the national level, on the resource policies and strategies of governments (Chapter 4)
- at the local level, on the needs of rural households, including their right and desire to participate in the crucial decisions that affect their livelihoods (Chapter 5).

The analyses in these chapters are intended to illuminate the main trade-offs among development goals and objectives and hence make explicit the choices confronting policy makers, planners and development agencies. However, before embarking on these analyses, we need to have a clear definition of sustainable agriculture and how it relates to other concepts of development. Thus, in Chapter 1 we describe briefly the way in which the notion of sustainability emerged out of the post-war evolution in ideas of development. In Chapter 2 we discuss the physical and biological basis of sustainability and the implications of adopting it as an indicator of agricultural performance.

We conclude the book in Chapter 6 with a summary of the priorities for the new phase of agricultural development which will carry us into the next century. We stress the need to:

- reorientate agricultural research and development efforts to cater for neglected marginal and resource-poor lands, as well as to improve the sustainability and stability of existing intensive agricultural production
- complement these research priorities with appropriate policy measures and institutional changes
- change existing philosophies and practices away from a top-down, technology-driven approach to one that is more sensitive to farmers' goals and needs.

At the local level we believe there is an important role for the analysis of changing agricultural conditions and systems through Rapid Rural Appraisal methods, and other techniques, which facilitate the flow of critical information to policy makers and planners at the national level. The Appendix provides an introduction to such rapid appraisal techniques.

Notes

1. For a discussion of the different strands of thinking in sustainable agriculture, see G K. Douglass, (ed.), *Agricultural Sustainability in a Changing World Order* (Boulder: Westview Press, 1984).

1. Ideas of Development

Although the overall aim of post-war development has been, and continues to be, the alleviation of mass poverty and the improvement in living conditions for the world's poorest populations, the strategy – and thus meaning – of economic development has undergone three important shifts over the post-war period.[1]

Economic growth

The first important phase, during the 1950s and early 1960s, equated economic development with economic growth, as defined by a sustained increase in real per capita gross national income.[2] In many ways the policies advocated and implemented during this phase were successful. Growth rates and savings grew, as did industrial capacity, in much of the Third World. Between 1950 and 1975 growth in gross domestic product (GDP) per capita averaged 3.4% per annum for all developing countries (Table 1.1). But critics argued that such growth did not necessarily "trickle down" to the poorest. Indeed there was accumulating evidence in the Third World of growing numbers of people below an absolute poverty line, of increasing income disparities between rich and poor, and of continuing underemployment and unemployment.[3] This realization led to two shifts in conventional, non-Marxist development thinking.

Growth with redistribution

The first shift – in the late 1960s and early 1970s – emphasized "growth with redistribution".[4] Economic growth was still the main objective, but the emphasis now was to be on growth that would improve the standard of living of the poorest income groups. Agriculture became the priority sector, since it had the potential to eliminate malnutrition and hunger, absorb surplus labour and boost foreign exchange earnings.[5] Export-led growth was also promoted, so stimulating the growth of labour-intensive manufacturing and providing foreign markets for commercial agriculture.[6]

Basic needs

A more recent and radical shift in perspective was the "basic needs" approach. Sparked by the call for a "basic needs strategy" at the 1976 World Employment Conference of the International Labour Organization (ILO), this approach argued that absolute poverty cannot be reduced unless the essential needs of the poor – nutrition, health, water supply, shelter, sanitation and education – are met, together with the fulfilment of certain non-material, but also

Table 1.1: Changes in gross domestic product (GDP) of developing countries, 1950–75

| | GDP per caput (1974 US$) | | Annual growth rate % p.a. |
	1950	1975	
South Asia	85	132	1.7
Africa	170	308	2.4
Latin America	495	944	2.6
East Asia	130	341	3.9
China	113	320	4.2
Middle East	460	1,660	5.2
All developing countries	160	375	3.4

Source: D. Morawetz, *Twenty Five Years of Economic Development 1950–75* (Baltimore, 1977), p.13.

Table 1.2: Developing country populations lacking basic needs

	1974		1982^c	
	Millions	*% total population*	*Millions*	*% total population*
Latin America	94	30.6	86	23.2
Near East^a	40	26.0	36	18.0
Asia^b	759	53.0	788	60.0
Tropical Africa	205	67.6	210	54.0
All developing countries	1,098	56.0	1,120	47.0

Notes: a   Middle East and African oil exporters
　　　 b   Excluding China
　　　 c   Projections from 1974 data

Source: M.J.D. Hopkins, "A global forecast of absolute poverty and employment", *International Labour Review*, vol. 119, pp.565–79, 1980.

essential, needs of self-reliance, security and cultural identity. Just how many people lack such basic needs is difficult to determine, but one estimate puts the figure at over a billion (Table 1.2). The basic needs strategy recognizes that growth by itself – even egalitarian growth or redistribution from growth – does not guarantee that basic needs will be met.[7] Instead, development policies must ensure these needs are met through increased supply of essential goods and services to the poor; through direct government intervention, if necessary, rather than relying on market forces. Moreover, this may have to entail some sacrifices in savings, productive investment and overall growth. The objective is a "new kind of economic growth", enabling basic needs to "be achieved by redistributing resources within the social sectors and by a reorientation of growth, so that the deprived participate".[8]

Sustainability

A concern for "sustainability" represents the most recent shift in development thinking.[9] In common with the "basic needs" strategy, the emphasis is on improving the livelihoods of the poor. However, this approach additionally argues that lasting improvement cannot occur in Third World countries unless the strategies which are being formulated and implemented are environmentally and socially sustainable; that is they maintain and enhance the natural and human resources upon which development depends.

This requires, on the one hand, national policies, regulations and incentives to induce economic behaviour that is "environmentally rational", i.e. that yields optimal benefits in both the short and long term from the world's endowment of natural renewable resources:[10] and on the other, development projects which are both ecologically sound and consistent with indigenous social values and institutions. To achieve this, it is argued, not only is local knowledge required but so is the full participation of the beneficiaries in the development process.

The green revolution

These shifts in overall development thinking have been mirrored by similar concerns within the somewhat narrower focus of agricultural development. Beginning in the 1950s there was an increasing preoccupation with the problem of feeding a rapidly growing world population. The goal of increasing per capita income was to be matched by rising per capita food production, and the means was the green revolution, largely funded by the international donor community and engineered by the International Agricultural Research Centres (IARCs). In essence it focused on three interrelated actions:

- breeding programmes for staple cereals that produced early maturing, day-length insensitive and high-yielding varieties (HYVs)
- the organization and distribution of packages of high pay-off inputs, such as fertilizers, pesticides and water regulation

- implementation of these technical innovations in the most favourable agroclimatic regions and for those classes of farmers with the best expectations of realizing the potential yields.[11]

Its impact in the Third World, particularly on wheat and rice production, has been phenomenal; between one-third and one-half of the rice areas in the developing world are planted with HYVs. In the eight Asian countries that produce ＇85 per cent of Asia's rice (Bangladesh, Burma, China, India, Indonesia, Philippines, Sri Lanka and Thailand) HYVs add 27 million tonnes annually to production, fertilizers another 29 million tonnes and irrigation 34 million tonnes.[12] Estimates of the contribution of new HYVs to increased wheat production in developing countries vary from 7 to 27 million tonnes.

Per capita food production in the developing countries has risen by 7% since the mid 1960s, with an increase of over 27% in Asia (Figure 1.1). Only in Africa has there been a decline.

Post green-revolution problems

These impressive results have been associated, though, with significant equity, stability and sustainability problems.[13] For instance, while producers have widely adopted the new HYVs irrespective of farm size and tenure, factors such as soil quality, access to irrigation water, and other biophysical-agroclimatic conditions have been formidable barriers to adoption. Farmers without access to the better-endowed environments have tended not to benefit from the new technologies, which partly accounts for the relative lack of impact of the green revolution in Africa. But even under favourable conditions in Asia or Latin America, a significant gap persists between performance on the agricultural research station and in the farmer's field.

The higher productivity of rice and wheat, relative to other crops for which no green revolution has yet occurred, has led many farmers to substitute these cereals for other staples and for more traditional mixed patterns of cropping. The resulting widescale monocropping has been associated with increased output variability, as crop yields grown with the new technologies may be more sensitive to year-to-year fluctuations in input use arising from shortages or price changes. For example, although the modern varieties are more responsive to fertilizers than traditional varieties, farmers have to apply higher levels of fertilizer to the modern varieties if they are to get the full benefits. Thus the gap between actual and potential yields is highly sensitive to the price of rice relative to that of fertilizer. Evidence from the Philippines, Thailand, Indonesia, Sri Lanka and Taiwan reveals that where it takes less than 0.8 kg of rice to buy 1 kg of urea, the yield gap attributed to fertilizer is 0.5 tonnes per hectare (tonnes/ha) or less. But where it takes over 1.5 kg of rice to buy 1 kg of urea, the yield gap generally exceeds 1 tonne/ha.[14]

Intensive monocropping with genotypically similar varieties has also led to increasing incidence of pest, disease and weed problems, sometimes aggravated by pesticide use. Severe outbreaks

Figure 1.1: Changes in per capita food production between 1964 and 1986

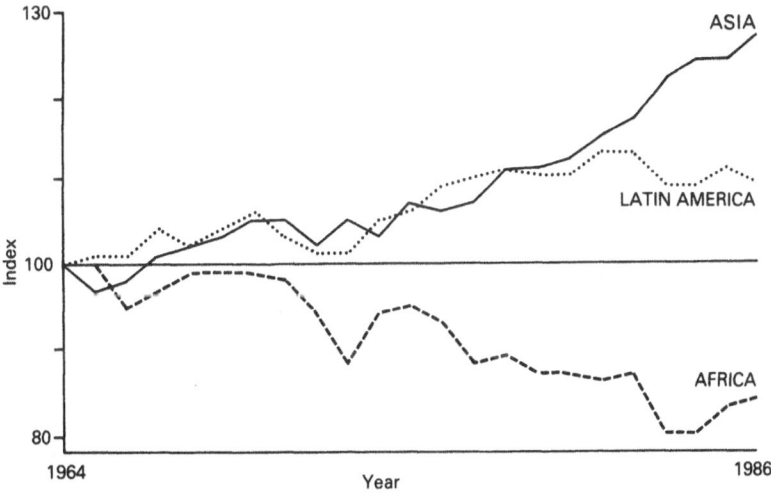

Sources: *FAO Production Yearbook* (Rome: Food and Agriculture Organization, various years)

of the brown planthopper occurred on rice in the 1970s with losses in 1977 in Indonesia of the order of 2 million tonnes. Planthoppers are naturally controlled by wolf-spiders and a variety of other natural predators and parasites which are destroyed by many of the pesticides commonly used on rice.[15]

There are now signs of diminishing returns to the HYVs and high pay-off inputs in intensive production. Perhaps, more important, the experience on less well-endowed farms, particularly in Africa, suggests there are real limits to replicating the successes of current green revolution technologies and packages in more marginal agricultural areas.

The problems, moreover, have not only been due to inappropriate technologies but to the nature of the accompanying national agricultural policies. These have tended to be short-term in nature, focusing exclusively on output growth and ignoring both the small farmer and the continuing degradation of the resource base. Credit, tenurial and marketing arrangements have tended to favour the adoption of the new technologies by larger rather than smaller farmers, while uniform pricing structures and standardized criteria for support services have encouraged inappropriate cropping patterns and their associated technologies. In these and many other respects such policies are diametrically at odds with the goal of sustainable development.

A new phase of agricultural development

These problems and failures are now commonly acknowledged by development practitioners and policy makers. In recent years they have also received prominence in a number of highly influential reports. Most notable has been the report of the World Commission on Environment and Development – the "Brundtland Report" – which argues for "environmentally sustainable economic growth" for the Third World and stresses that, "although the agricultural resources and the technology needed to feed growing populations are available", global food security requires "increasing food production to keep pace with demand while retaining the essential ecological integrity of production systems."[16]

The arguments in favour of promoting a more sustainable development approach, particularly the dismantling of policies and incentive structures that stand in its way, are also slowly being accepted by the international donor community.[17] A recent review by the World Bank of renewable resource management in its agricultural projects concluded that there must be three criteria for "successful" agricultural development: "First, it must be sustainable, by insuring the conservation and proper use of renewable resources. Second, it must promote economic efficiency. Third, its benefits must be distributed equitably."[18]

The theme has been further endorsed by the Consultative Group on International Agricultural Research (CGIAR), which in a recent report calls attention to the technological and research priorities required for making agricultural production in the Third World more sustainable. Similarly, the International Fund for Agricultural Development (IFAD) recently held a consultative meeting to discuss strategies for implementing sustainable agricultural development in resource-poor environments and spreading benefits to the rural poor.[19] And the Asian and Near East Bureau of the US Agency of International Development (USAID) has produced a report on sustainable agriculture as part of its overall commitment to a comprehensive Environmental and Natural Resources Strategy.[20]

In sum, the evolution of development thinking is now pointing to a post green-revolution phase characterized by the term "sustainable agriculture". But to understand fully the implications of this, we have to be clear what is implied by the adoption of sustainability as an indicator of agricultural performance. This issue is the focus of the next chapter.

Notes

1. For a more detailed overview, see Edward B. Barbier, "The concept of sustainable economic development", *Environmental Conservation*, vol.14, no.2 (Summer 1987).
2. Influential works during this phase included Walt W. Rostow, *Stages of Economic Growth* (New York: Cambridge University Press, 1960); Ragnar Nurske, *Problems of Capital Formation in Underdeveloped*

Countries (New York: Oxford University Press, 1953); and Paul Rosenstein-Rodan, "Problems of industrialisation of Eastern and South-Eastern Europe", *Economic Journal* (June–September 1943).

3. See, for example, Hollis Chenery *et al.*, *Redistribution with Growth* (New York: Oxford University Press for the World Bank, 1974); and Hollis Chenery and Moses Syrquin, *Patterns of Development 1950–1970* (New York: Oxford University Press for the World Bank, 1975).

4. Influential works included Irma Adelman, "Development economics – a reassessment of goals", *American Economic Review*, vol.65, no.2 (1975); Chenery *et al.*, op. cit.; Gunnar Myrdal, *Asian Drama: An inquiry into the poverty of nations*, 3 vols (New York: Pantheon, 1968); and Dudley Seers, "What are we trying to measure?", *Journal of Development Studies*, vol.8, no.3 (1972).

5. See, for example, Bruce F. Johnston and Peter Kilby, *Agriculture and Structural Transformation: Economic strategies in late-developing countries* (London: Oxford University Press, 1975); and T.W. Schultz, *Transforming Traditional Agriculture* (New Haven: Yale University Press, 1964).

6. See, for example, Gerald K. Helleiner, *International Trade and Economic Development* (Harmondsworth: Penguin Books, 1972); I.M.D. Little, Tibor Scitovsky and M.F.G. Scott, *Industry and Trade in Some Developing Countries* (London: Oxford University Press, 1970); and Hla Myint, *The Economics of Developing Countries* (London: Hutchinson, 5th edn, 1980).

7. Paul Streeten *et al.*, *First Things First: Meeting basic needs in developing countries* (New York: Oxford University Press for the World Bank, 1981), p.108. See also ILO, *Employment, Growth and Basic Needs, a One World Problem* (Geneva: ILO, 1976); and Frances Stewart, *Planning to Meet Basic Needs in Developing Countries* (London: Macmillan, 1985).

8. Stewart, op. cit., p.211.

9. World Commission on Environment and Development, *Our Common Future* (Oxford: Oxford University Press, 1987).

10. Barbier, op.cit.

11. Gordon R. Conway, *Helping Poor Farmers – A Review of Foundation Activities in Farming Systems and Agroecosystems Research and Development* (New York: Ford Foundation, 1987), p.3.

12. Per Pinstrup-Andersen and Peter B.R. Hazell, "The impact of the green revolution and prospects for the future", in J. Price Gittinger, Joanne Leslie and Caroline Hoisington (eds), *Food Policy:*

Integrating supply, distribution, and consumption (Baltimore: Johns Hopkins University Press for the World Bank, 1987), p.107.

13. For further discussion of other post green-revolution problems, see Conway, *Helping Poor Farmers*, op. cit., ch.1; Pinstrup-Andersen and Hazell, op.cit.; Mohammed Alauddin and Clem Tisdel, "Population growth, new technology and sustainable food production: the Bangladesh case", paper prepared for the Fourth World Congress of Social Economics, Toronto, Canada, 13–15 August 1986; Randolph Barker, Robert W. Herdt and Beth Rose, *The Rice Economy of Asia* (Washington: Resources for the Future, 1985); Gordon R. Conway, "Rural resource conflicts in the UK and Third World – issues for research policy", *Papers in Science, Technology and Public Policy*, no.6 (London: Imperial College of Science and Technology, 1984); Gordon R. Conway and David S. McCauley, "Intensifying tropical agriculture: the Indonesian experience", *Nature*, vol.302 (1983), pp.228–89; and KEPAS, *The Sustainability of Agricultural Intensification in Indonesia: A report of two workshops of the Research Group on Agro-Ecosystems* (Jakarta: KEPAS, 1983).

14. Randolph Barker, Robert W. Herdt and Beth Rose, *The Rice Economy of Asia* (Washington, DC: Resources for the Future, 1985), p.227. The relationship is particularly striking in the dry season, probably because the wet-season yield-gaps embody a greater degree of variability caused by weather, insects and disease than do the dry-season gaps.

15. Conway and McCauley, op. cit.; Christopher Joyce, "Nature helps Indonesia cut its pesticide bill", *New Scientist* (June 16 1988), p.35; Peter E. Kenmore, *Ecology and outbreaks of a tropical pest of the green revolution, the brown planthopper, Nilaparvata lugens Stal.* (Berkeley: University of California); A.G. Cook and T.J. Perfect, "The population characteristics of the brown planthopper, *Nilaparvata lugens*, in the Philippines", *Ecological Entomology*, vol.14 (1989), pp.1–9.

16. World Commission on Environment and Development, op. cit.

17. Recent key publications are T.J. Davies and I.A. Schirmer (eds), *Sustainability Issues in Agricultural Development*, Proceedings of Seventh Agriculture Sector Symposium (Washington, DC: World Bank, 1987); World Commission on Environment and Development, *Food 2000: Global policies for sustainable agriculture* (London: Zed Books, 1987); and CGIAR Technical Advisory Committee, *Sustainable Agricultural Production: Implications for international agricultural research* (Rome: TAC Secretariat, FAO).

18. World Bank, *Renewable Resource Management in Agriculture* (Washington, DC: Operations Evaluation Department, World Bank, 1988), p.iv.
19. The background paper for that meeting drew from Edward B. Barbier, "Sustainable agricultural development, the environment and the role of marginal farmers and pastoralists: the policy framework", paper prepared for the International Consultation on Environment, Sustainable Development and the Role of Small Farmers, Sponsored by IFAD (Rome, 11–13 October 1988).
20. Edward B. Barbier, *Economic Aspects of Sustainable Agriculture: A strategy for Asia and the Near East*, paper prepared for the Environment and Natural Resources Strategy Project of the Asia and Near East Bureau (Washington, DC: USAID, February 1989). See also a companion piece, Theodore Panayotou, *Natural Resource Management: Strategies for sustainable agriculture in the 1990s* (Cambridge, Mass: Harvard Institute for International Development, August 1988).

2. Indicators of Agricultural Performance

Chapter 1 traced the major post-war trends in development think-ing and, in particular, the recent emergence of the concept of sustainable economic development. This coincided with, and indeed in part grew out of, a search for a new, post green-revolution approach to agriculture in the Third World, which would place greater emphasis on sustainability and equity. In this chapter we intend to examine more closely what "sustainability" implies, both in theory and practice, and how it relates to other measures of agricultural performance.

The basis of sustainability

Clearly sustainability has to be viewed within the context of the overall agricultural production system (Figure 2.1). Agricul-ture depends for its success on the exploitation of natural and human-made resources, using human skills and labour. The out-puts are products in the form of food or fibre and their produc-tion, together with that of non-agricultural goods and services, helps to secure both national economies and the livelihoods of individual households.

The first issue to be addressed is the sustainability of the resource base.

The sustainability of resources

The conventional distinction between non-renewable and renew-able resources has obvious implications for sustainability. Resour-ces such as fossil fuels, which drive farm machines and produce agrochemicals, are intrinsically exhaustible and hence their use cannot be indefinitely sustained. We are consequently faced with two questions. The first is: what is the "best" balance between present and future consumption? In the context of agriculture the issue is: what proportion of non-renewable resources should be devoted to current food production as compared with a perhaps greater need in the future? Needless to say, this is a question that is extremely difficult to answer, but it has to be addressed. We do have a fairly good estimate of the likely population growth and its demand on food for the next 50, perhaps 100 years. The imponderables are future technological innovations and the extent to which they can relieve the pressure on non-renewable resources.

The second question is somewhat easier to answer: how can the benefits from current exploitation of non-renewable resources be sustained? One general answer is to invest the profits so as to provide a sustainable, long-term return. Such investment

Figure 2.1: The basic elements of the agricultural production system

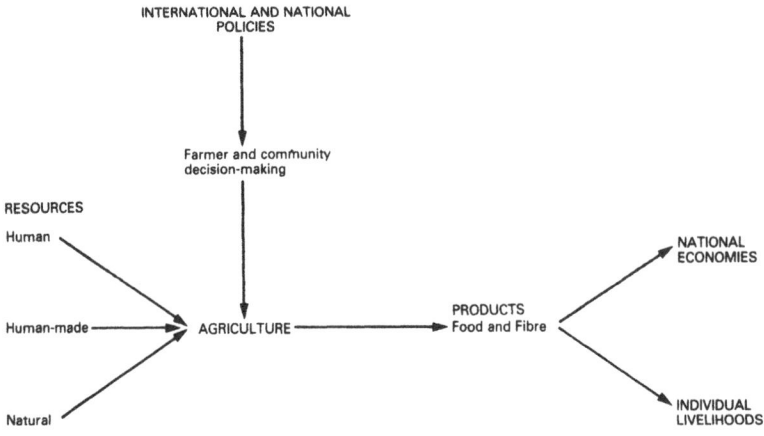

Note: There are, of course, numerous complications to this simple picture, including feedback loops of many kinds between the elements.

may be in industrial technologies, or in human skills, or in technologies for sustainable renewable-resource exploitation. In the case of oil-rich countries, such as Indonesia, it can be argued that using oil revenues to subsidize fertilizers, although providing immediate benefit, for example in the attainment of rice self-sufficiency, does not create a capacity for a continuing return. Investment of the revenues in the rehabilitation of irrigation schemes, on the other hand, generates an intrinsically sustainable development, and in this sense is to be preferred.

Renewable resources

Besides solar radiation, which to all intents and purposes is inexhaustible, most natural resources on which agriculture depends are potentially renewable. They include the soil and its nutrients; water, which is derived directly or indirectly from rainfall; and the diversity of wildlife; together with a great variety of ecological processes, including the capacity of the environment to control pests or assimilate wastes.

It is, of course, possible to treat these as exhaustible resources, i.e. as sources of profit for investment in other productive activities. Forests can be felled and the revenues used for agricultural development, or agricultural land given over to urban and industrial growth. However, strong arguments can be made for insisting that such resources are managed so as to derive the returns from their intrinsic ability to renew themselves. In the case of land and soil fertility, it is possible to envisage future food production systems in highly artificial environments, for example concentrated hydroponic systems that utilize nitrogen and power produced via nuclear fusion energy. But the doubts surrounding the viability of such schemes suggest that it would be imprudent to destroy the existing natural resource endowment now, and rely on such a future being affordable or even feasible. The prudent strategy is to husband renewable resources in such a way as to provide a long-term sustainable base for production.

Destruction of renewable resources is especially characteristic of frontier societies, as prevalent in North America in the seventeenth and eighteenth centuries as in, say, Brazil or Indonesia today. The common perception among all levels of such societies is that natural resources are essentially available in unlimited amounts. Soil is allowed to erode, soil fertility is destroyed, fuel-wood is exhausted, as is the assimilative capacity of the environment for wastes, without fully appreciating that limits are approaching. In most developed countries, these limits have been superseded or overcome, so far, through the intensive application of capital and technology and through the importation of new materials and foodstuffs from the rest of the world. Developing countries, however, are recognizing that they cannot afford the technological investment, nor do they have dependent countries which they can exploit. At the same time in the developed countries it is becoming increasingly clear that many of the technological solutions, for example use of pesticides and artificial fertilizers or "industrial" livestock production, have high and unexpected costs and, more importantly, are themselves in many respects not sustainable. Substituting a technological input for a renewable resource does not necessarily improve sustainability.

Global estimates for the current losses of renewable natural resources are very crude and open to considerable argument.

Table 2.1: On-site soil losses due to erosion in Java (all figures rounded)

Type of land use	Land area 10^6 ha	Soil loss 10^6 tonnes	Soil loss tonnes/ha
Sawah (wet rice land)	4.6	2	0.5
Forest	2.4	14	5.8
Degraded forest	0.4	35	87.2
Wetlands	0.1	–	–
Tegal (rainfed cropland)	5.3	737	138.3
Total	**12.9**	**787**	**61.2**

Source: W.B. Magrath and P. Arens, *The Costs of Soil Erosion on Java – A Natural Resource Accounting Approach* (Washington, DC: World Resources Institute, 1987).

But somewhat better estimates are available for national losses. One example is a recent study on losses due to erosion in Java and its impact on agricultural production (Tables 2.1 and 2.2). The heaviest losses occur on sloping upland fields planted to rice, cassava or other food crops, and although there are considerable off-site costs arising from the need to maintain and dredge lowland reservoirs, irrigation systems and harbours, the main consequences are the lost yields in upland agriculture itself.

Mismatched technologies

Destruction of renewable resources, however, is not the only form of resource mismanagement. Renewable resources may be wasted if they are subject to inappropriate technologies. Waste can be said to occur if the potential benefits of naturally sustainable processes are not fully realized, through lack of knowledge or appropriate skills. Technological or labour inputs are commonly used in place of these processes with the consequence that the costs of production are frequently higher than they might otherwise be. In Citanduy, West Java, for example, the cost of bench terracing is approximately US$560–2,075 per hectare (/ha) (1979 prices) and involves from 750 to over 1,800 person-days of work.

Table 2.2: Summary of erosion costs in Java

	$m per annum
On-site	324
Off-site:	
Irrigation system	8–13
Harbour costs	1– 3
Reservoirs	16–75
Total	**349–415**

Source: W.B. Magrath and P. Arens, *The Costs of Soil Erosion on Java – A Natural Resource Accounting Approach* (Washington, DC: World Resources Institute, 1987).

Yet this expenditure is largely to reinstate the intrinsic productivity of the land.[1] The capacity of vegetation to prevent erosion is in many respects a free good, and terracing is thus a cost which is both unnecessary and, by contrast with the natural resource capacity, inherently less sustainable.

Similarly, pesticides are a costly replacement for natural control mechanisms. When farmers were trained in Integrated Pest Management (IPM) techniques for control of brown planthopper and other pests of rice, they were able to reduce their insecticide sprays from over four to less than one in a growing season and their average rice yields rose from 6.1 to 7.4 tonnes/ha.[2]

Mismatching of technologies is particularly apparent where technological packages are applied on a large scale, in the belief that the natural resources they are intended to manage are uniform. Such a package approach was appropriate during the green revolution when the aim was to disseminate a limited number of high-yielding rice varieties and their accompanying inputs throughout a relatively uniform expanse of irrigated lowland. But this approach is highly inappropriate for the the development of the more marginal uplands in the tropics. In Java the uplands consist of a great diversity of agro-ecosystems which differ not only in bio-physical but in socio-economic terms. Any uniform package, whether of terracing or cropping patterns, is likely in some places to be unsuitable and non-sustainable. Thus, recommendations for terracing, while appropriate for volcanic soils, are frequently disastrous on limestone soils, causing even greater erosion than before.

Mismatched technologies may also have deleterious effects which extend well beyond the agricultural production system itself. Agrochemicals, for example, may not only undermine the natural agricultural resource-base but, if improperly used, may destroy natural resources over a wide area, affecting other resource-based activities and, in certain circumstances, causing human disease and death. As far as agriculture is concerned the sustainability of agrochemical use is then threatened by increasing regulation of these compounds to protect human health and the wider environment.

The evidence for pesticides causing harm to humans in the

developing countries is incomplete but there are sufficient anecdotal accounts to suggest that pesticide-related illnesses and deaths are seriously under-reported. The symptoms of pesticide poisoning are frequently confused with cardio-vascular and respiratory diseases, or with epilepsy, brain tumours and strokes. A study of human mortality in the Philippine rice-growing regions of Luzon, where pesticide use has grown dramatically in recent years, found a highly significant correlation between increasing death rates of rural men and women, and increasing pesticide use.[3] Moreover the mortalities were highest at the peak time of spraying. The evidence is clearly circumstantial, but if correct it would imply many thousands of deaths a year are resulting from pesticide use in the Philippines.

Even less well known are the adverse effects of nitrogen fertilizers in the developing countries, although the developed countries have begun to introduce restrictions because of health hazards.[4] The most likely hazard is methaemoglobinaemia or the "blue baby syndrome", which particularly affects infants in the first few months of life. It is associated with high levels of nitrate in drinking water which is also contaminated with bacteria. The other risk is cancer, particularly gastric cancer. This is common in some developing countries, for example parts of Chile and Colombia, although there is at present no clear link with the use of nitrogen fertilizer. The link is stronger in the incidence of bladder cancer in the intensive agricultural lands of the Nile delta of Egypt. There a strong correlation exists between cancer, nitrate and bacterial levels in the water, and the incidence of the parasite disease schistosomiasis.

There is also evidence that such technologies, and indeed agricultural development itself, may be having an adverse effect on the global environment as a whole and, in particular, its capacity to provide for a continuance of our present climate and to furnish a shield against damaging solar radiation.[5] Global temperatures appear to be rising as a result of the production of various gases, notably carbon dioxide and nitrous oxide, that create the so-called "greenhouse effect". Agricultural development is partly responsible in that it indirectly contributes to the clearing and burning of forests. More directly, nitrogen fertilizer use results in emissions of nitrous oxide to the atmosphere and although, at present, fertilizers are estimated to account for between only 1–4% of emissions, the proportion is likely to increase in the future. There is also some theoretical evidence to suggest that increasing nitrous oxide may be reducing the ozone layer in the stratosphere and hence removing some of the protection against cancer-inducing radiation.

Any major increase in global temperature or reduction in the ozone layer would have profound effects on agriculture and development in general. As yet there is still insufficient information to predict what will happen, or to ascribe causes. If nitrogen fertilizers are more closely implicated, then restrictions on their use may eventually be required.

Internal and external resources

One way of establishing, in principle, whether particular agricultural production systems are likely to be inherently sustainable, is to consider the local community or individual farm and to classify the available resources into those which are "internal" and "external"[6] (Table 2.3).

Internal resources are the resources available within the farm or community and immediate environment. They include rainfall, biologically fixed nitrogen, nutrients from lower soil strata and biological pest control based on indigenous natural enemies. They are inherently renewable and thus have the potential to be used on a "sustained" basis, indefinitely, through ecologically-sound methods of farming.

In contrast, external resources have to be obtained from outside the farm or community, and include irrigation water from a distant source, synthetic nitrogen fertilizer or phosphates, and chemical pesticides. Most non-renewable resources used in agricultural production, such as fossil fuels and their by-products, are external resources. Consequently, the dependence on external resources which are not provided or obtained "free of charge" means that the farming household must generate a surplus of production, cash or something else of value, to exchange for the external resource. Moreover, the cost of

Table 2.3: Agricultural production resources which are derived from internal and external sources

Internal resources	*External resources*
Sun – source of energy for plant photosynthesis	**Artificial lights** – used in greenhouse food production
Water – rain and/or small, local irrigation schemes	**Water** – large dams, centralized distribution, deep wells
Nitrogen – fixed from air, recycled in soil organic matter	**Nitrogen** – primarily from applied synthetic fertilizer
Other nutrients – from soil reserves recycled in cropping system	**Other nutrients** – mined, processed and imported
Weed and pest control – biological, cultural and mechanical	**Weed and pest control** – chemical herbicides and insecticides
Seed – varieties produced on-farm	**Seed** – hybrids or certified varieties purchased annually
Machinery – built and maintained on farm or in community	**Machinery** – purchased and replaced frequently
Labour – most work done by the family living on the farm	**Labour** – most work done by hired labour
Capital – source is family and community, reinvested locally	**Capital** – external indebtedness, benefits leave community
Management – information from farmers and local community	**Management** – from input suppliers, crop consultants

Source: C.A. Francis and J.A. King, "Cropping systems based on farm-derived, renewable resources", *Agricultural Systems*, vol.27 (1988), pp.67–75.

acquiring the resource, and often its supply, are not directly controllable by the household or even a farming community. It is also important to note that even where there is no direct monetary cost to obtaining the resource, there may be a real cost incurred by the household. The most obvious examples are the time allocated for obtaining seeds and other inputs from local distribution centres, or for searching for fuelwood and fodder from distant areas, or for carrying water from the nearest tubewell.

Dependence on external resources is not only frequently costly; it also tends to make the production system more vulnerable to external stresses and shocks, such as changes in the costs and supply of these resources. Of longer term significance, such dependence may also lead to fundamental changes in the farming system that make it more vulnerable to the vagaries of the local environment. This is one explanation of the failure of the adoption of green revolution "packages" of hybrid seeds, fertilizers and pesticides in resource-poor environments. Such packages are often less suited to these environments compared to the lower yielding, yet better adapted, internal resources used in traditional farming systems.

Labour, capital, machinery and management can be either internal or external to the farm. When these resources are primarily internal – for example a family-owned and operated farm – then households have a greater degree of control over decisions concerning the allocation of resources and their long-term management. As an example, a recent study of land tenure in the hills of Nepal indicates that production is directly related to the degree of control.[7] It is highest on land which is cultivated by farmers who own the land and lowest on lands tilled by informal tenants on a contract basis; where a farmer both owns and rents land, production is higher on the former. More importantly, landowners who cultivate their own land or participate in the management of land rented to informal tenants, have a greater incentive to manage it sustainably. In contrast, informal tenants are unlikely, on their own, to take an interest in the long-term productivity of the land they are working.

A definition of agricultural sustainability

So far we have been using the term sustainability in a way that is roughly equivalent to persistence or durability. The implicit question that is asked about a particular agricultural practice or system is: will it last? Will it be productive not only in the immediate future, but over the long term, for present and future generations? Durability, however, has to be assessed in terms of the forces that are likely to cause the agricultural practice or system to collapse. We thus need a definition that embraces these forces.

The common usage of the word "sustainable" suggests an ability to maintain some activity in the face of stress – for example to sustain physical exercise, such as jogging or doing press ups – and this seems to us also the most technically acceptable meaning. We thus define agricultural sustainability as the ability to maintain productivity, whether of a field or farm or nation, in the face of stress or shock.[8] A stress may be increasing salinity, or erosion, or debt; each is a frequent, sometimes continuous, relatively small, predictable force having a large cumulative effect. A major event such as a new pest or a rare drought or a sudden massive increase in input prices would constitute a shock, i.e. a force that was relatively large and unpredictable. Following stress or a shock the productivity of an agricultural system may be unaffected, or may fall and then return to the previous level or trend, or settle to a new lower level, or the system may collapse altogether.

Sustainability thus determines the persistence or durability of a system's productivity under known or possible conditions. It is a function of the intrinsic characteristics of the system, of the nature and strength of the stresses and shocks to which it is subject, and of the human inputs which may be introduced to counter these stresses and shocks.

The biophysical subsidy, often in the form of a fertilizer application, and intended to counter the stress of repeated harvesting, is a ubiquitous input. Sustainability is maintained only by renewed fertilizer application. Another common form of input is a control agent; for example, a pesticide to counter pest

Figure 2.2: Contrasting dynamics of pesticide and biological methods for the control of pests

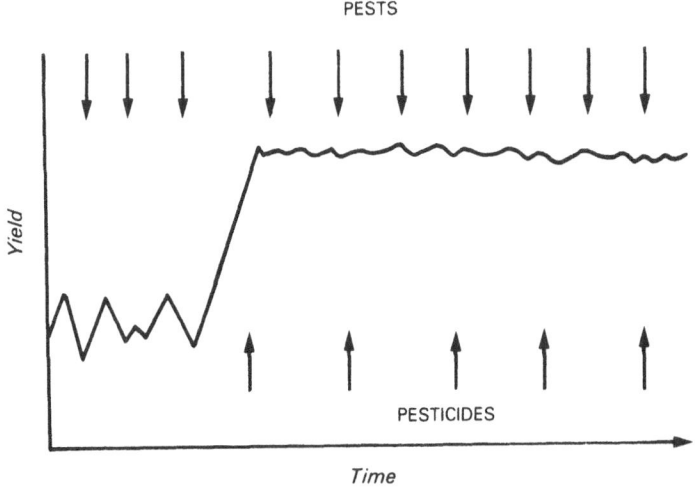

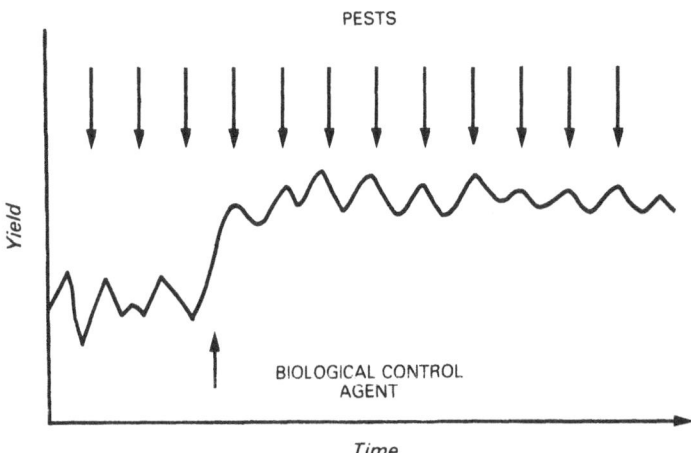

or disease attack. Again, sustainability may necessitate repeated pesticide applications, but an alternative strategy may be the introduction of a biological control agent, such as a parasitic wasp, which may so permanently alter the intrinsic sustainability characteristics of the system as to obviate the need for further intervention (Figure 2.2). This also illustrates the process of building sustainability into a system, i.e. substituting internal resources for external resources. Controlling pests by pesticides can be sustainable, providing the pesticides are affordable and used selectively. However, the value of introducing a biological control agent is that it exploits a renewable natural resource – the parasite or predator – and is hence relatively cheap and inherently a sustainable process. Box 2.1 lists a number of examples of sustainable agricultural technologies.

Productivity, stability and equitability

Sustainability, however, is clearly not the only criterion by which we judge agricultural development or even development as a whole (Figure 2.3). Productivity is the most commonly used measure of agricultural performance, but also important is the stability of production, from month to month and year to year, and the manner in which that production is shared, i.e. its equitability.

Productivity
We define productivity as the output of valued product per unit of resource input. The three basic resource inputs are land, labour and capital. Strictly speaking, energy is subsumed under land (solar energy), labour (human energy) and capital (fossil fuel energy). Similarly, technological inputs, such as fertilizers and pesticides, are components of capital, but both energy and technology can be treated, for many purposes, as separate inputs.

Common measures of productivity are yield or income per hectare, or total production of goods and services per household or nation; but a large number of different measures are possible, depending on the nature of the product and the resources being

Figure 2.3: Indicators of agricultural performance

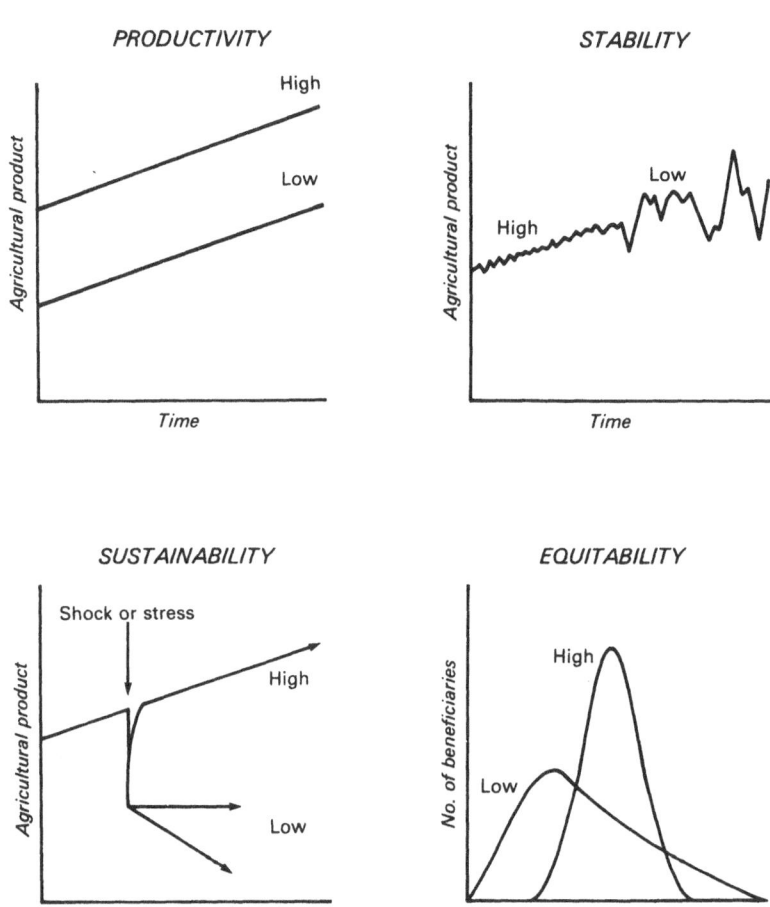

Source: Gordon R. Conway, *Helping Poor Farmers – a Review of Foundation Activities in Farming Systems and Agroecosystems Research and Development* (New York: Ford Foundation, 1987).

BOX 2.1: EXAMPLES OF AGRICULTURAL TECH-
NOLOGIES THAT HAVE A HIGH POTENTIAL
SUSTAINABILITY

Intercropping – the growing of two or more crops simultaneously
on the same piece of land. Benefits arise because crops exploit
different resources, or mutually interact with one another. If one
crop is a legume it may provide nutrients for the other. The interactions may also serve to control pests and weeds.

Rotations – the growing of two or more crops in sequence on the
same piece of land. Benefits are similar to those arising from intercropping.

Agroforestry – a form of intercropping in which annual herbaceous
crops are grown interspersed with perennial trees or shrubs. The
deeper-rooted trees can often exploit water and nutrients not available to the herbs. The trees may also provide shade and mulch,
while the ground cover of herbs reduces weeds and prevents erosion.

Sylvo-pasture – similar to agroforestry, but combining trees with
grassland and other fodder species on which livestock graze. The
mixture of browse, grass and herbs often supports mixed livestock.

Green manuring – the growing of legumes and other plants in
order to fix nitrogen and then incorporating them in the soil for
the following crop. Commonly used green manures are *Sesbania*,
and the fern *Azolla* which contains nitrogen-fixing blue-green algae.

Conservation tillage – systems of minimum tillage or no tillage,
in which the seed is placed directly in the soil with little or no
preparatory cultivation. This reduces the amount of soil distur-
bance and so lessens run-off and loss of sediments and nutrients.

Biological control – the use of natural enemies, parasites or predators, to control pests. If the pest is exotic these enemies may be
imported from the country of origin of the pest; if indigenous,
various techniques are used to augment the numbers of the existing
natural enemies.

Integrated pest management – the use of all appropriate techniques
of controlling pests in an integrated manner that enhances rather
than destroys natural controls. If pesticides are part of the programme, they are used sparingly and selectively so as not to interfere
with natural enemies.

considered. Yield may be in terms of kilograms of grain, tubers, leaves, meat or fish, or any other consumable or marketable product. Alternatively, yield may be converted to value in calories, proteins or vitamins, or to its monetary value at the market. In the latter case it is measured as income as a function of expenditure, i.e. profit. But, frequently, the valued product may not be the yield in conventional agricultural terms. It may be employment generation, or an item of amenity or aesthetic value; or one of a wide range of products that contribute, in ways that are difficult to measure, to social, psychological and spiritual well-being.

Stability
Stability may be defined as the constancy of productivity in the face of small disturbing forces arising from the normal fluctuations and cycles in the surrounding environment. Included in the environment are those physical, biological, social and economic variables that lie outside the agroecosystem under consideration. The fluctuations, for example, may be in the climate or in the market demand for agricultural products. Productivity may be defined in any of the ways described above and its stability measured by, for example, the coefficient of variation in productivity, determined from a time series of productivity measurements. Since productivity may be level, or rising or falling, stability will refer to the variability about a trend.

Equitability
Equitability is defined as the evenness of distribution of the productivity of the agricultural system among the human beneficiaries, i.e. the level of equity that is generated. Once again, the productivity may be measured in many ways, but, commonly, equitability will refer to the distribution of the total production of goods and services, or the net income of the agroecosystem under consideration, i.e. the field, farm village or nation. The human beneficiaries may be the farm household, or the members of a village or a national population.

Equitability may be measured by a Lorenz curve, Gini coefficient or some other related index. In practice, though, it is difficult to define equitability in a purely positive sense, as measures available reflect different value judgements. Thus, equitability is often the evenness of distribution of productivity among the human beneficiaries, according to need.

In most situations equitability is affected not only by the distribution of products but also by the distribution of costs. That is, equitability refers to the distribution of net benefits. In many cases, as we will argue later, productivity involves significant external costs and these have to be included in the computation of equitability.

Trade-offs

Defined in this way, the three key indicators are fairly readily understandable by all concerned in development, whether they be policy makers, project designers and implementers, or the farmers themselves. Furthermore, when the indicators are viewed as normative goals, the trade-offs between them are similarly clear and understandable. They occur equally for farmers in their day-to-day decisions and for nations determining agricultural strategies and policies.

Such trade-offs are not new phenomena. Both the Sumerians in the arid lands of ancient Mesopotamia and the Maya in the tropical forests of Central America appear to have sacrificed sustainability in the quest for higher productivity. And there are examples in history where equitability and sustainability were achieved at the expense of productivity – for example in the manorial agriculture of medieval Europe. More recently, these trade-offs are recognizable in the development policies that have been pursued in the Third World over the past 40 years.

There are numerous examples. Large-scale irrigation projects can increase productivity but at the expense of sustainability and equitability. Similarly, too much emphasis on equitability can inhibit productivity. Such trade-offs may even be involved in the adoption of apparently wholly benign resource-conserving technologies. For instance, pest control using a biological control agent may well be more sustainable, yet the farmer may have to accept a lower and more fluctuating yield (Figure 2.2).

Productivity, efficiency and sustainability

The complexities involved in such trade-offs are well illustrated if we consider the concept of efficiency. For biologists and agronomists, efficiency is consistent with the broad definition of productivity used here, i.e. output over input, where the outputs and inputs are measured in physical or biological units.[9] Maximum efficiency (or productivity) then occurs either when output alone, or the output per unit of input, is maximized (Box 2.2).

However, economists distinguish between this definition of efficiency, which they refer to as technical efficiency, and economic efficiency, which is also consistent with the broad definition of productivity, except the inputs and outputs are defined in monetary terms.[10] Maximum economic efficiency is equivalent to maximum profit and lies somewhere between the two technically efficient points shown in Figure 2.4. The simple system described in this figure can, of course, be expanded to multiple inputs and multiple products, but the principle for calculating efficiencies remains the same.

Having clarified these concepts, the trade-off question now is whether economic efficiency is compatible with sustainability. Some say it is. For example, the World Bank study, cited in Chapter 1, states: "There is no conflict between efficiency and sustainability, i.e. any use of renewable resources which leads to the exhaustion of those resources cannot be efficient."[11] This is obviously true if we are referring to technical efficiency – if total output or output per unit input declines then both efficiency and sustainability fall. However, in terms of economic efficiency the statement, without important qualifications, is not necessarily true. The economics literature on "optimal extinction"

BOX 2.2: THE CONCEPT OF EFFICIENCY

If, for simplicity, we consider a single agricultural input, say nitrogen fertilizer, then the output–input relationship can be described as in Figure 2.4.

Figure 2.4: Measures of efficiency for a single input–output agricultural system

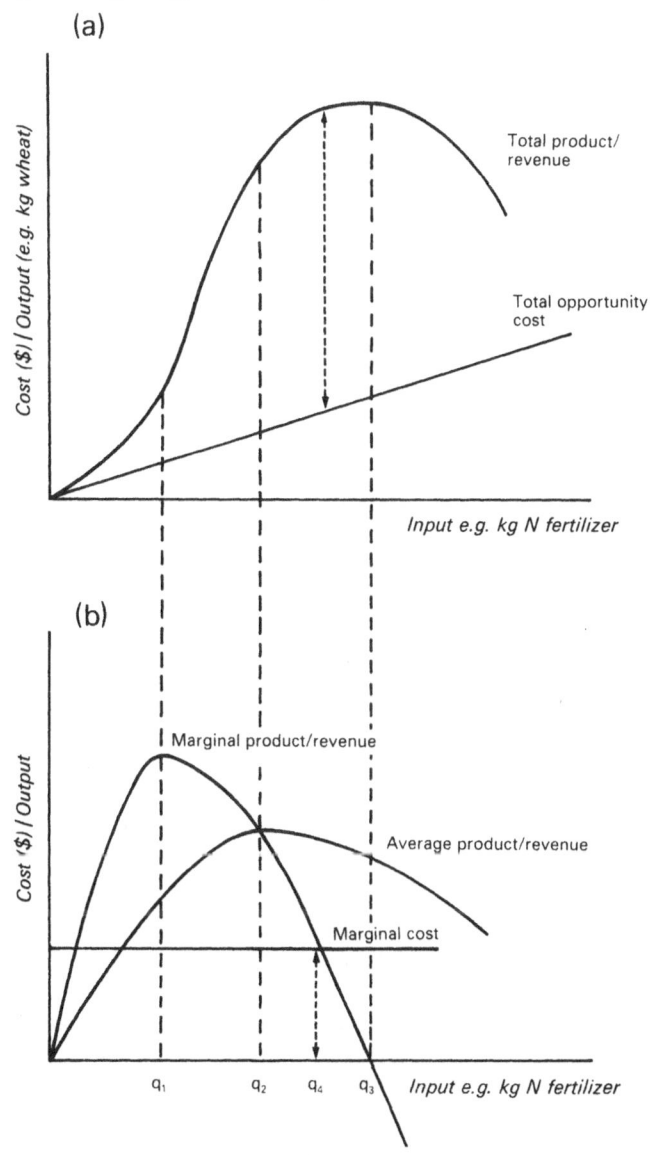

It is assumed that total production follows an S-shaped curve as described here, although other shapes are possible and common. The marginal product is the rate of change of the total product. It increases in the early growth phase of the production curve to a maximum (q_1), but then falls away to zero (q_3) when total production is maximized. The average product is simply the total product divided by the input. Total revenue, total cost and marginal cost refer to similar measures of the value of the products and the opportunity cost of the input in monetary terms – where the opportunity cost is the value of the input in its next best alternative use or function. All of these are different, useful measures of productivity.

Productivity is regarded as technically efficient either when the average product is maximized (q_2) or when total product is at its peak (q_3). However, economic efficiency occurs somewhere between these two points. If we translate total product in terms of its value, in US$ for example, to total revenue, and add the total opportunity cost of the input, profit will be maximized when the difference between these two is greatest, i.e. when marginal revenue equals marginal opportunity cost (q_4). This is the point of economic efficiency.

of renewable resources suggests that under certain conditions economic efficiency is wholly consistent with a renewable resource being exhausted within a limited period of time.[12] Crucial factors in this are the dynamics of the agricultural production system over time and the alternative avenues for investment.

Underlying the simple production diagram in Figure 2.4(a) are a variety of natural renewable resources on which production depends, in addition to the fertilizer input. One such will be soil quality. The question this raises is *whether the soil quality will be maintained for year after year for the levels of input and output that produce maximum profit*. In terms of Figure 2.4(a), will the soil quality be regenerated after the crop is harvested so that in the following year the production curve is as it was before? If the soil does not regenerate sufficiently for this to happen, then important choices have to be made.

Figure 2.5 depicts possible scenarios for the production curve from year to year. In (a) the level of input that produces

Figure 2.5: Effects of level of input on the production curve from year to year

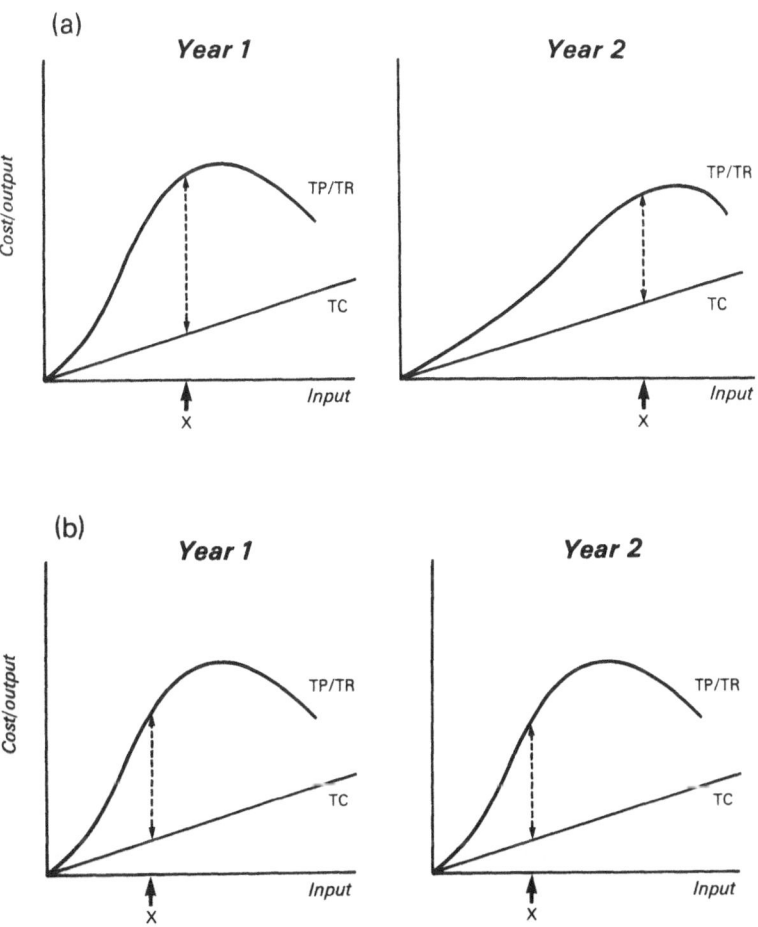

Note: X indicates level of input chosen. In (a) X equals level giving maximum profit; in (b) X is well below maximum profit level.

maximum profit results in damage to soil quality that is not fully recovered by the following year. To gain maximum profit requires a higher level of input. In (b) by choosing a consistently lower level of input from the beginning the system is maintained, although the level of profit is less than is possible in the first year under (a). In both these cases there is a trade-off between economic efficiency and sustainability. Which course the investor or farmer chooses to follow depends on the level of profit compared with other alternative investments. If the level of profit in situation (b) is better than an alternative investment then this is the course the investor will adopt. But often the rate of recovery or regeneration of the natural resource produces a sustainable level of profit which is too low, and then it may pay the investor to follow course (a) – maximize profit each year regardless of the damage being done to the resource, take the profits and invest them elsewhere. (This, of course, is what happened in the whaling industry. The natural population growth rate of whales is typically less than 5%, i.e. less than most other returns on money invested.) The second course may also be one that small farmers pursue, in order to satisfy desperate short-term basic needs, in the hope that there will be an alternative to the exhausted resource in the future.

It may also pay investors and farmers to follow course (a) if the costs are very small, or the value of the product is very high; i.e. in Figure 2.4 the difference between the total revenue and costs curves is very large. In this situation it may be a long time before there is any significant decline in profits, or the short-term profits may be such as to constitute a large capital gain.

In summary the conditions for pursuing efficiency at the expense of sustainability will be:

(1) if the regenerative capacity of the resource is low enough and the future is heavily discounted, then it is economically efficient to exhaust the resource. That is, higher discounted net returns are obtained through exhausting the resource as quickly as possible and investing the proceeds in other assets, whose value will increase much faster; *and*

2) equivalently, if the cost of production is low enough, or the value of each unit of product is high enough, then it may also

be economically efficient to exhaust the resource quickly.

Nevertheless even if these conditions exist, the trade-off is likely to be bounded by other considerations. For instance, bounds can be set through some criterion of sustainability, such as the need to preserve a minimum stock of the resource for future generations, possibly for uses which are as yet unconceived. The authors of the World Bank statement (referred to earlier) presumably believed that the overwhelming dependence of agriculture in the Third World on renewable resources makes it unlikely that the above conditions hold for these resources, or that the bounds need to be very tightly set. That is, degradation and depletion of renewable resources impose such high costs on agricultural development that economic efficiency in both short term and long term is bound to be impaired.

Agroecosystems

As the above discussion clearly demonstrates, the trade-offs involved in agricultural development are often highly complex and embrace a wide range of factors, ecological and social as well as economic. This complexity presents a considerable challenge both in terms of analysis and in the practical implementation of development projects. An answer lies in using systems frameworks that help make the key processes and factors explicit. One such conceptual system is the agroecosystem.

So far we have referred to agricultural systems, but in essence these are ecological systems transformed by human action to produce food and fibre. In this transformation, the great diversity of wildlife in the original ecological systems is reduced to a restricted assemblage of crops, pests and weeds. Take for example a ricefield (Figure 2.6). There is a strengthening of the biophysical boundary of the system; a bund is created around the ricefield, for example. The basic renewable ecological processes still remain: competition between the rice and weeds, herbivory of the rice by pests and predation of pests by their natural enemies. But these are now overlaid and regulated by agricultural processes of cultivation, subsidy (with fertilizers), control (of water, pests and diseases),

Figure 2.6: The ricefields as an agroecosystem

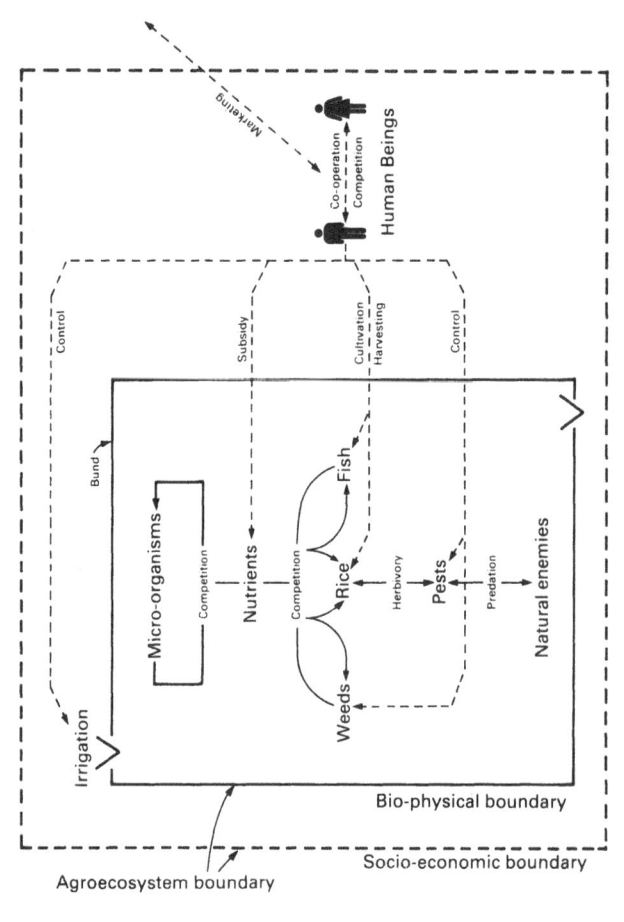

Source: Gordon R. Conway, "The properties of agroecosystems", *Agricultural Systems*, vol. 24, no. 2 (1987), pp. 95-117.

Figure 2.7: The hierarchy of agroecosystems

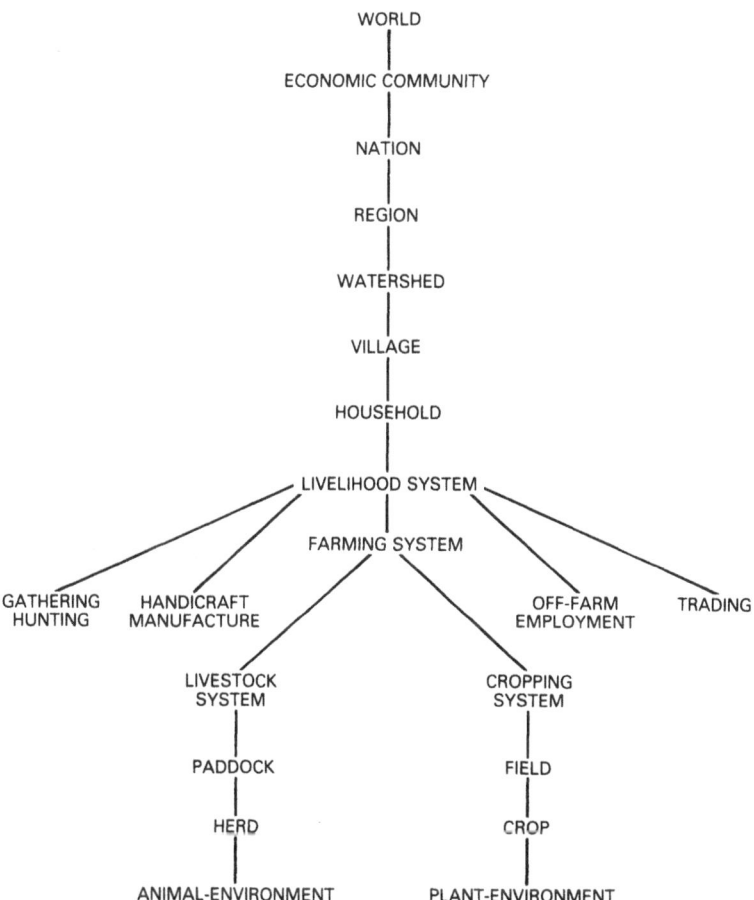

Source: Gordon R. Conway, "The properties of agroecosystems", *Agricultural Systems*, vol.24, no.2 (1987), pp.95–117.

harvesting and marketing. Dominating the system are human goals and the consequences of human social and economic co-operation and competition. As a result the system is as much a socio-economic system as it is an ecological system, and has both biophysical and socio-economic boundaries. It is this new complex agro-socio-economic-ecological system, bounded in several dimensions, that we call an agroecosystem. Within it the trade-offs between productivity, stability, sustainability and equitability occur.

Hierarchies

The most widely recognized agroecosystem is the crop field conceptualized in Figure 2.6, or its analogue – the livestock paddock. But if agroecosystems are defined so as to include both ecological and socio-economic components, then we can envisage a classical hierarchy of such systems (Figure 2.7). At the bottom of the hierarchy is the agroecosystem comprising the individual plant or animal, its immediate micro-environment, and the people who tend and harvest it. Examples where this exists as a recognizably distinct system are the lone fruit tree in a farmer's garden, or the milk cow in a stall. The next level is the field or paddock; the hierarchy continues upwards in this way, each agroecosystem forming a component of the agroecosystem at the next level. Near the top is the national agro-ecosystem composed of regional agroecosystems linked by national markets, and above that the world agroecosystem consisting of national agroecosystems linked by international trade. The higher up the hierarchy the greater is the apparent dominance of socio-economic processes, but ecological processes remain important and, at least in sustainability terms, crucial to achieving human goals. It may seem to be overextending definitions to regard the nation as an agroecosystem but we believe such a conceptualization is essential if the key trade-offs are to be explicitly recognized and analysed.

It is also important to appreciate that the behaviour of higher systems in such a hierarchy is not readily discovered simply from a study of lower systems, and vice versa. This has consequences not only for analysis but for agricultural policy and planning. It follows that each level in the agroecosystem hierarchy has to be analysed, and developed both in its own right and in relation to the other levels above and below, and this totality of understanding used as the basis of development. (Methods for carrying out such analysis are described in the Appendix.)

Trade-offs in the hierarchy
Furthermore, trade-offs do not only occur within agroecosystems, but also between agroecosystems in the hierarchy. Thus for a farm, high stability and sustainability may depend on a complementary diversity of crop fields and livestock systems, each of which produces less than its maximal potential, is more variable in yield, and individually less sustainable than is the total farm. A similar situation can occur between the nation and its agricultural regions.

Perhaps the most important of such trade-offs occurs between the productivities of individual farms and that of the nation as a whole. Here again economists use the term efficiency and distinguish *private efficiency* – the efficiency of the production system from the point of view of its users, and *social efficiency* – how the production system affects the allocation of resources to society as a whole.

For the individual farm, as we have seen, economic efficiency is attained by maximizing the discounted *net* private returns, i.e. the benefits the farmer receives less the costs he or she incurs in producing those benefits. However, for society as a whole, efficiency refers to the maximization of the discounted net *social* returns, i.e. the benefits less the costs accruing not just to the farmer but to all individuals who are affected by the farmer's actions on his or her farm.

Economists further qualify this goal of maximizing net returns by also requiring that it meet the criterion of *Pareto optimality*, i.e. that it is not possible to further change the allocation of resources without making someone worse off. Thus a system of resource use is regarded as being inefficient if it is still possible to re-allocate resources and make some people better off while making no one else worse off. Note that this introduces an element of equity into the definition of efficiency, although of course achieving Pareto optimality does not necessarily mean an increase in the evenness of distribution of net benefits; it simply ensures that no one is worse off.

A simple example can illustrate the relationship between private efficiency, social efficiency and sustainability. Let us take a simple upland farm agroecosystem producing an annual crop such as cassava for a single household's subsistence and income needs.[13] We assume it is a low-input system, i.e. the household cannot afford or gain access to modern inputs such as inorganic fertilizers. Suppose that production from this system can be sustained indefinitely, except for the environmental stress imposed on the system from prolonged soil erosion leading to declining soil fertility. As a result of this stress, future cassava yields will decline and the system may collapse. The farmer thus incurs what economists refer to as the user costs of soil erosion – the loss of future soil productivity through the erosion caused by current use of the resource for crop production. Such user costs are part of the overall private costs that the household attempts to minimize in its quest for efficiency in production. Under normal conditions, one would expect that the household would find these user costs so significant that it would have to bring soil erosion under control in order to maximize its discounted net returns. In such instances, the pursuit of private efficiency will also ensure the overall sustainability of the agricultural production system.[13]

But there are also circumstances – leading to conditions 1 and 2 referred to on p.48 – under which the household may ignore the user cost of soil erosion in its drive for production efficiency. For example, the lack of secure tenure, or open access to forests that can be converted to agriculture, may make the household less concerned about the future productivity of the land on which it is currently growing cassava. Alternatively, some upland soils, such as those based on limestone, may be very poor in quality and have a low regenerative capacity. Under such conditions, the household may find that its discounted net returns are higher not from controlling soil erosion, but through exhausting the soil as quickly as possible in order to maximize current yields. As condition 2 indicates, this will also be the case if the production cost of cassava is low, or if the price of cassava is high. As will be discussed in Chapter 4, more often than not these costs and prices are influenced by government policies, such the use of input subsidies and procurement policies to increase the producer price of food.

However, even under conditions where the pursuit of production efficiency by a farming household also ensures the sustainability of production, it does not automatically follow that this outcome is also socially efficient. The existence of *external costs*, costs imposed on other individuals who do not receive compensation or a share of the benefits of the production system, may be a factor. For example, supposing that in order to reduce the user cost of soil erosion to zero, which happens to be consistent with efficient cassava production, our farming household would only have to reduce the rate of erosion to 10 tonne/ha per year. If this were true of all upland farming households, then upland production systems would be efficient and sustainable. Unfortunately, though, the impact of an annual erosion rate of 10 tonnes/ha in the uplands might be sedimentation of irrigation canals downstream. The result is a loss of productivity experienced by lowland irrigated farmers, which is the external cost of upland soil erosion. From society's perspective, since individuals – the lowland farmers – are being made worse off, this situation is not (Pareto) optimal. Moreover, it could threaten the sustainability of lowland production. It would be more socially efficient to find some means of compensating upland farmers to reduce their erosion rates further in order to eliminate the external downstream costs to lowland farmers. If such a solution were found, then social efficiency and the sustainability of lowland, as well as upland, production would be complementary.

Once again, though, a socially optimal solution might be found that does not necessarily ensure sustainability of production. For example, society might find that a less costly alternative to compensating upland farmers to reduce erosion further may be to provide affected lowland farmers with off-farm employment opportunities as their yields start declining. Under this scenario, water supplies and irrigation facilities will be allowed to collapse, ending the sustainability of lowland production. But from a social perspective, this loss of agricultural sustainability is not crucial to the maximization of overall net returns. We are back to a situation where conditions 1 and 2 are in force.

In summary there are frequently trade-offs between private and social efficiency and between both of these and sustainability.

Such trade-offs can be minimized but only after very detailed and systematic analysis, and through the development of carefully targeted policies.

Short- and long-term equitability

We have argued in this chapter that the pursuit of economic efficiency in agriculture, under some conditions, may allow the exhaustion of those resources important for sustainability. Even the pursuit of social efficiency does not necessarily lead to the sustainability of natural resources essential to agricultural production. But in addition to these conflicts, the issue of equitability further complicates the picture. As we have already pointed out, attaining Pareto optimality does not necessarily improve the distribution of net benefits in society. There will commonly be trade-offs between social efficiency and equitability, in both the short term and long term.

Thus a *potential* Pareto improvement may exist, but it is not realized in the short term. In the example of downstream sedimentation described above, lowland farmers might benefit in the long run from the least-cost solution of providing off-farm employment opportunities. This would be an efficient solution because it offers the chance of a Pareto improvement. However, in the short term, lowland farmers might have difficulty in adjusting to the new employment conditions; for example in acquiring the necessary skills – for construction or factory work, small-scale trading, cottage industries and so on – and their income may thus suffer initially. Clearly there is then a short-term trade-off between social efficiency and equitability.

Of greater significance, though, is a long-term inter-generational equity consideration. Some economists argue that the conservation of essential natural resources can be justified on the grounds of ensuring equal access to these resources for future generations so that they, too, can achieve sustainable and secure livelihoods. This argument particularly applies to those resource-poor farmers and pastoralists who are directly dependent on the resource base for their livelihoods and for whom there is little alternative means of income and employment in the near future. It is also relevant to low and lower-middle income countries whose agricultural development is dependent on successful exploitation of existing renewable resources. In these circumstances conserving essential resources may be a viable development goal, even if under some conditions it may lead to outcomes not wholly consistent with the objective of economic efficiency.[15]

In the next three chapters we look further at the nature of these trade-offs from a hierarchical perspective, focusing first on the international constraints to sustainable agriculture in developing countries, second on the national policies and strategies required by these countries to improve sustainability, and finally on the conditions necessary to secure sustainable livelihoods for individual households.

Notes

1. The source of these estimates is R. Bernstein and R. Sinaga, "Economics", *Composite Report of the Watershed Assessment Team* (Jakarta: USAID, 1983), Technical Appendix VI. These estimates, do not include the additional costs to the farmer of periodic maintenance of terraces, waterways and drop structures.
2. Christopher Joyce, "Nature helps Indonesia cut its pesticide bill", *New Scientist* (16 June 1988).
3. See Michael E. Loevinsohn, "Insecticide use and increased mortality in rural central Luzon, Philippines", *Lancet* (13 June 1987), pp.1359–62; and Jennifer A. McCracken and Gordon R. Conway, *Pesticide Hazards in the Third World: New Evidence from the Philippines* (London: IIED, 1987).
4. See Gordon R. Conway and Jules N. Pretty, "Fertiliser risks in the developing countries", *Nature*, vol.334 (1988), pp.207–8; Gordon R. Conway and Jules N. Pretty, *Fertiliser Risks in the Developing Countries: A review* (London: International Institute for Environment and Development, 1988); Jules N. Pretty and Gordon R. Conway, *Cancer Risk and Nitrogen Fertilisers: Evidence from developing countries* (London: IIED, 1988); Jules N. Pretty and Gordon R. Conway, *The Blue-baby Syndrome and Nitrogen Fertilisers: A high risk in the tropics* (London: IIED, 1988).
5. See Gordon R. Conway and Jules N. Pretty, *Agriculture as a Global Polluter* (London: IIED, 1989).
6. See also Charles A. Francis, *Internal Resources for Sustainable Agriculture* (London: Sustainable Agriculture Programme, IIED, 1988)

on which the following definitions are based.

7. Integrated Development Systems (IDS), *The Land Tenure System in Nepal* (Kathmandu: IDS, 1986).

8. See Gordon R. Conway, "The properties of agroecosystems", *Agricultural Systems*, vol.24, no.2 (1987), pp.95–117.

9. See C.R.W. Spedding, J.M. Walsingham and A.M. Hoxey (eds), *Biological Efficiency in Agriculture* (London: Academic Press, 1981).

10. For a good introduction to the basic economic concepts, see Christopher Ritson, *Agricultural Economics: Principles and policy* (Oxford: BSP Professional Books, 1988).

11. World Bank, *Renewable Resource Management in Agriculture* (Washington, DC: Operations Evaluation Department, World Bank, 1988), p.iv.

12. See, for example, the discussions of Colin W. Clark, *Mathematical Bioeconomics: The optimal management of renewable resources* (New York: John Wiley and Sons, 1976); Partha S. Dasgupta, *The Control of Resources* (Oxford: Basil Blackwell, 1982); Anthony C. Fisher, *Resource and Environmental Economics* (Cambridge: Cambridge University Press, 1981); and Vernon L. Smith, "Control theory applied to natural and environmental resources: an exposition", *Journal of Environmental Economics and Management*, vol.4 (1977), pp.1–24.

13. For a complete case study of the response of upland farming households to soil erosion see Edward B. Barbier, *The Economics of Farm-level Adoption of Soil Conservation Measures in the Uplands of Java*, World Bank, Working Paper no.11, (Washington, DC: Environment Department, 1988).

14. See Talbot Page, *Conservation and Economic Efficiency: An approach to materials policy* (Baltimore: Johns Hopkins University Press, 1977); and David W. Pearce, "Foundations of an ecological economics", *Ecological Modelling* (July 1986).

15. For example, in David W. Pearce, Edward B. Barbier and Anil Markandya, *Sustainable Development and Cost Benefit and Cost Benefit Analysis*, LEEC Paper 88–03 (London: London Environmental Economics Centre, 1988), it is shown that sustainability can be introduced into cost-benefit analysis by setting a constraint on the depletion and degradation of the stock of natural capital. However, the result is to produce a level of economic activity for a portfolio of projects that is different from the strict Pareto optimal level.

3. International Constraints

The sustainability of agricultural development in the developing countries is crucially dependent on international relationships and world trade. Declining commodity prices and terms of trade – coupled with problems of debt, exchange rate and financial instability – are serious constraints to orderly or rapid development. But before we discuss these, we need to underline the importance of the fundamental stress on developing economies placed by growing populations. This factor greatly compounds the effects of adverse international relationships.

Population and food demand

The populations of developing countries are still predominantly rural. On average, 62% of the labour force in developing economies is engaged in agriculture and other primary-resource based activities such as forestry, fishing and hunting. These activities contribute to an estimated 20% of the gross domestic product (GDP). In low-income economies, as much as 72% of the labour force is in agriculture and related activities, accounting for 32% of GDP.[1]

For the future, this pattern is likely to change only slowly. The greater part of global population increase will take place in the Third World, the 1985 population of 3.7 billion increasing to perhaps 6.8 billion by 2025. Although by the first decade of the next century rural populations in most developing countries will start declining, they will continue to increase in some of the poorest countries. Assuming no change in the distribution of land and other resource assets, the number of subsistence farmers, pastoralists and landless people – groups that represent three-quarters of the agricultural households in developing economies – will increase to nearly 220 million households or some 1 billion people, by the year 2000.[2]

As a result of these trends, between 1980 and 2000 increased total food demand is projected to exceed the growth of food output in all developing regions except Asia, despite increases in per capita food production in Latin America, North Africa and the Middle East. Even though Latin America is expected to have the highest food production per capita growth rate, demand is projected to increase even faster. Because of its rapid population growth, Sub-Saharan Africa's food consumption is estimated to grow 3.6% a year, substantially outpacing the projected growth in food output. Indeed, per capita food production in the region is expected to continue to decline.[3]

Food security

Inevitably, many developing regions will continue to be dependent on food imports, and in some instances external assistance, to meet domestic consumption requirements. Providing grain production continues to grow in the industrialized countries and there is a major improvement in Soviet agriculture, global food supplies may keep pace with global demand. The lack of food security in the developing countries – defined as the access by all people at all times to enough food for an active, healthy life – will arise from a lack of purchasing power on the part of nations and households rather than from inadequate global food supplies.[4] The disturbing facts are that in recent years food insecurity has become even worse in many developing countries, notwithstanding higher per capita food production. Moreover, despite record levels of world food production and excess supplies, about 730 million people in developing countries do not obtain enough energy from their diet to allow them to have an active working life. About two-thirds of the undernourished live in South Asia and a fifth in Sub-Saharan Africa; four-fifths of the undernourished live in countries with very low average incomes.[5]

The roots of most solutions to overcoming chronic food security will have to be sought at the national level. They include policies to ensure sustainable increases in food production in low-income, food-deficient developing countries; improved distribution systems; reductions in population growth; and the alleviation of poverty, particularly guaranteeing secure and sustainable livelihoods for vulnerable groups. But there are also serious constraints at the international level which are equally crucial and need to be resolved. They include the poor purchasing power of developing countries and their lack of financing for food purchases, together with numerous physical distribution problems – such as the desired size and location of global reserve stocks, methods of sharing storage costs, and of acquiring and releasing stocks to minimize disruptions in importer and exporter nations.

Perhaps more important are the distorting effects of the current pattern of global agricultural production. At present large export subsidies in the United States and other major agricultural exporters are having a strong negative impact on the agriculture of importing developing nations. Such subsidies depress prices in importing economies, thus reducing the incentives for domestic farmers to expand production. Over the long run, agricultural production and food security suffer, especially in the poorest countries.[6]

Vulnerability to external shocks and stresses

These trends represent only some of the factors working against the sustainability of developing economies. Because individual developing countries, even the larger ones, have so little control over their external environment they are highly vulnerable to a range of external stresses and shocks. They must take as given important international economic factors, such as the growth of world markets, protectionism, terms of trade, cost and availability of foreign credit and capital, aid flows and so forth.[7] Two types of external economic stress or shock are important – those arising from adverse developments in world agricultural markets and trade, and those from adverse developments in the world economy as a whole.

Table 3.1: Low income economies with high export concentration in predominantly agricultural commodities[a]

	Contribution of 33 main commodities to total exports	Main export commodities	
		1	2
over 90%			
Burundi ($230)	98.5	coffee (91.2)	cotton (2.8)
Uganda ($230)[b]	98.0	coffee (94.0)	cotton (1.8)
Equatorial Guinea	95.4	cocoa (71.5)	timber (18.5)
Rwanda ($280)	94.7	coffee (66.6)	tin (17.0)
Malawi ($170)	91.9	tobacco (49.8)	sugar (19.8)
Cuba	90.2	sugar (88.5)	tobacco (0.8)
over 80%			
Burma ($190)	81.2	rice (43.2)	timber (29.0)
over 70%			
Ethiopia ($110)	71.7	coffee (61.5)	hides & skins (6.8)
over 60%			
Chad	65.1	cotton (60.7)	hides & skins (4.5)
Nepal ($160)	63.5	rice (26.0)	hides & skins (16.9)
Central Afr. Rep. ($260)	63.2	coffee (28.7)	timber (25.4)
Tanzania ($290)	60.0	coffee (29.8)	cotton (13.3)
over 50%			
Benin ($260)	50.8	cotton (20.7)	cocoa (14.2)
Burkina Faso ($150)	50.6	cotton (45.0)	hides & skins (4.0)
Vanuatu	50.6	copra (38.4)	cocoa (4.4)

Notes
a   Calculated in terms of percentage contributions to the value of total merchandise exports in 1981–83. U.S. dollar figure after each country listed indicates GNP per capita in 1985. Low-income economies are those with GNP per person of $400 or less in 1985.
b   GNP per capita in 1984.
Sources: World Bank, *Commodity Trade and Price Trends* (Washington, DC: World Bank, 1986); and World Bank, *World Development Report* (Washington, DC: World Bank, 1986 and 1987).

Table 3.2: Lower middle income economies with high export concentration in predominantly agricultural commodities[a]

	Contribution of 33 main commodities to total exports	Main export commodities	
		1	2
over 70%			
Guyana ($580)[b]	76.6	sugar (34.4)	bauxite (29.4)
Nicaragua ($770)	74.0	coffee (28.5)	cotton (23.9)
Honduras ($720)	71.8	bananas (28.2)	coffee (22.7)
over 60%			
El Salvador ($820)	67.3	coffee (56.5)	cotton (7.0)
Ivory Coast ($660)	67.1	cocoa (24.2)	coffee (19.4)
Mauritius ($1,090)	61.8	sugar (59.9)	tea (1.9)
Paraguay ($860)	60.9	cotton (37.0)	timber (17.7)
Costa Rica ($1,300)	60.6	bananas (25.2)	coffee (25.0)
over 50%			
Colombia ($1,320)	59.9	coffee (49.2)	bananas (4.6)
Dominican Republic ($790)	58.3	sugar (38.0)	coffee (9.1)
Guatemala ($1,250)	50.5	coffee (28.9)	cotton (6.6)

Notes
a   Calculated in terms of percentage contributions to the value of total merchandise exports in 1981–83. US dollar figure after each country listed indicates GNP per capita in 1985. Lower-middle income economies are those with GNP per person of $1,600 or less in 1985. Note that no country with GNP per capita greater than $1,600 in 1985 had 50% or more of its exports comprised of agricultural commodities.
b   GNP per capita in 1984.
Sources: World Bank, *Commodity Trade and Price Trends* (Washington, DC: World Bank, 1986); and World Bank, *World Development Report* (Washington, DC: World Bank, 1986 and 1987).

Table 3.3: Debt and debt service ratios in predominantly agricultural exporting low-income economies[a]

| | External public debt[b] as percentage of GNP | | Debt service as percentage of: | | | |
| | | | GNP | | Exports | |
	1970	1985	1970	1985	1970	1985
over 90%						
Burundi (98.5)	3.1	39.7	0.3	2.0	2.3	16.6
Uganda (98.0)	7.5	–	0.4	–	2.9	–
Eq. Guinea (95.4)	–	–	–	–	–	–
Rwanda (94.7)	0.9	19.1	0.1	0.9	1.2	4.3
Malawi (91.9)	44.2	75.7	2.2	7.4	7.7	–
Cuba (90.2)	–	–	–	–	–	–
over 80%						
Burma (81.2)	5.0	42.1	1.0	2.8	17.2	51.4
over 70%						
Ethiopia (71.7)	9.5	37.1	1.2	2.2	11.4	10.9
over 60%						
Chad (65.1)	9.9	–	0.9	–	4.2	–
Nepal (63.5)	0.3	22.5	0.3	0.5	–	4.0
Central Afr. Rep. (63.2)	13.5	44.9	1.7	2.0	5.1	11.8
Tanzania (60.0)	20.1	48.5	1.3	1.0	5.2	16.7
over 50%						
Benin (50.8)	15.2	66.9	0.6	2.2	2.3	–
Burkina Faso (50.6)	6.6	46.4	0.7	2.5	6.8	–
Vanuatu (50.6)	–	–	–	–	–	–

Notes
a   Percentage figure after each country listed indicates contribution of 33 main primary commodities to total exports as indicated in Table 3.1. Low-income economies are those with GNP per person of $400 or less in 1985.
b   External public debt outstanding and disbursed.
– = figures not available.
Sources: World Bank, *Commodity Trade and Price Trends* (Washington, DC: World Bank, 1986); and World Bank, *World Development Report* (Washington, DC: World Bank, 1987).

Adverse developments in world agriculture

Many low and lower-middle income developing economies are highly dependent on predominantly agricultural export earnings, often from one or two major commodities (see Tables 3.1 and 3.2). Moreover they are increasingly dependent on such earnings to service their rising external debt (see Tables 3.3 and 3.4). Such trends are reinforced by the structural adjustment policy packages advocated by the International Monetary Fund (IMF) and World Bank. These encourage developing countries to reorientate their economies away from the production of non-tradable goods and services, towards export commodities as a means of paying their debts. The effect has been a growing reliance on the expansion of agricultural and other primary commodity exports.

Yet real agricultural commodity prices have exhibited a long-term historical decline since 1950, falling to record lows over the period 1984–86 (see Figure 3.1). Between the fourth quarter of 1983 and the second quarter of 1986, the current dollar index for agricultural commodities fell by 13%. The greatest declines were in fats and oils, non food agricultural commodities, and cereals.

There are several factors underlying these trends:[8]

- *trade subsidies* – since the mid-1980s, export subsidies by the United States and European Economic Community (EEC) have risen dramatically as each has tried to protect domestic agricultural producers and increase its respective share of world trade
- *domestic subsidies* – farm income and price supports in the United States and EEC also contribute to excess global supplies of some agricultural commodities, and to lower prices
- *structural imbalances* – these factors are compounded by the likely continuation of a substantial excess production capacity in US agriculture well into the mid-1990s and by the failure of the EEC to control overproduction of certain commodities
- *protectionism* – lower global agricultural prices have increased the pressure in industrialized countries for import restrictions and other measures to protect domestic producers from international competition. In the short term, as contraction in production lags behind, these restrictions coupled with export subsidies further depress world prices.

Figure 3.1: Real agricultural commodity prices, 1950-86

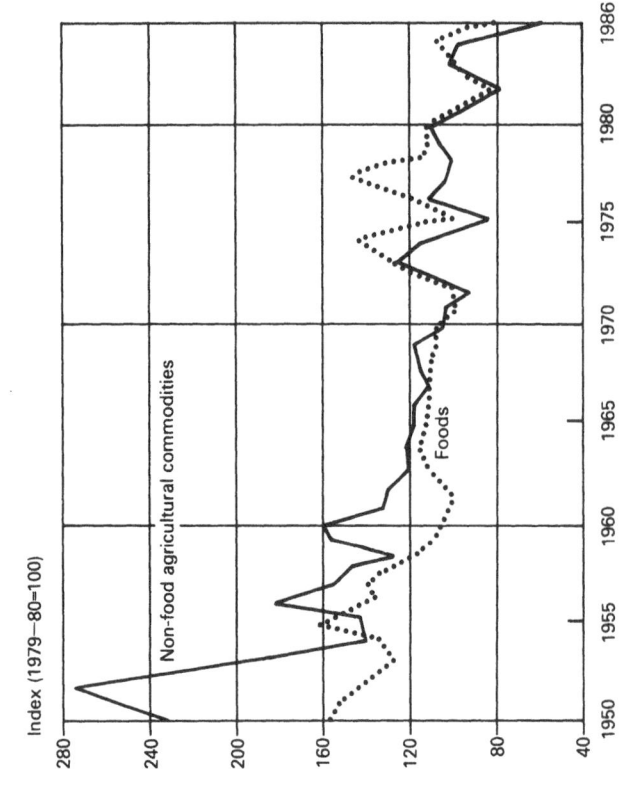

Source: World Bank, *World Development Report 1987* (Washington, DC: World Bank, 1987), Figure 2.3.

Despite lower prices, many low-income food-deficient countries lack the purchasing power to acquire the developed country surpluses, and if they are able to benefit this is likely to be short-lived since their farmers then have less incentive, over the long term, to increase domestic production to overcome chronic food shortages. Lower prices and long-term loss of external markets also mean that agricultural exporting countries, especially those poorer nations dependent on agriculture for a substantial share of export earnings (see Tables 3.1 and 3.2), will continue to suffer. Even in the short term, because many food-importing developing countries also export agricultural products, the net positive balance of payments effect of the global agricultural price decline is likely to be negligible.

Adverse developments in the world economy
The stresses and shocks produced by the dynamics of international economic relations are, in many respects, even more pervasive and significant for sustainable agricultural development. They arise from a range of interrelated factors.

SLOWER GROWTH AND TRADE
Since 1984, the peak year of the brief recovery that began in 1982, both global economic growth and trade have slowed significantly. Over the last ten years, the pre-1973 tendency for trade to grow faster than gross domestic product has disappeared. As a result, global demand for agricultural raw materials from developing countries has been weak, which has put further pressure on commodity prices.

DEBT PROBLEMS
The increasing debt-servicing obligations of developing countries have placed them under great pressure to restrain imports and expand exports. For many economies, this has meant a radical restructuring of agriculture towards export markets and an increasing share of export earnings allocated to debt servicing (see Tables 3.3 and 3.4). Although these pressures have been somewhat reduced – by lower real costs of borrowing and the lower costs of energy and raw materials – many developing-nation loans are short term in nature. Increasing their repayment period and accepting often austere structural adjustment policies may be the only long-term solution. But this means that external debt will remain a

Table 3.4: Debt and debt service ratios in predominantly agricultural exporting lower-middle income economies[a]

| | External public debt[b] as percentage of GNP | | Debt service as percentage of: | | | |
| | | | GNP | | Exports | |
	1970	1985	1970	1985	1970	1985
over 70%						
Guyana (76.6)	–	–	–	–	–	–
Nicaragua (74.0)	19.5	185.2	3.0	1.6	10.5	–
Honduras (71.8)	13.6	68.8	0.9	5.4	3.1	17.6
over 60%						
El Salvador (67.3)	8.6	39.6	0.9	5.3	3.6	16.3
Ivory Coast (67.1)	18.8	88.5	2.9	9.0	7.0	17.4
Mauritius (61.8)	14.7	39.8	1.4	6.6	3.2	11.5
Paraguay (60.9)	19.2	55.8	1.8	5.6	11.8	12.9
Costa Rica (60.6)	13.8	105.1	2.9	13.3	10.0	36.6
over 50%						
Colombia (59.9)	18.5	28.5	1.7	4.3	12.0	29.2
Dominican Rep. (58.3)	14.5	58.6	0.8	5.1	4.4	16.1
Guatemala (50.5)	5.7	19.8	1.4	2.3	7.4	21.3

Notes
a   Percentage figure after each country listed indicates contribution of 33 main primary commodities to total exports as indicated in Table 3.2. Lower-middle income economies are those with GNP per person of $1,600 or less in 1985.
b   External public debt outstanding and disbursed.
– = figures not available.

Sources: World Bank, *Commodity Trade and Price Trends* (Washington, DC: World Bank, 1986); and World Bank, *World Development Report* (Washington, DC: World Bank, 1987).

persistent and long-term stress on already vulnerable economies and agricultural systems.

EXCHANGE RATE INSTABILITY

The 1980s have seen large swings in the value of the US dollar. It appreciated nominally and in real terms in the early 1980s but then with the emergence and persistence of the United States' current account deficit and the surpluses of Japan and West Germany, this trend was rapidly reversed against other major trading currencies. 1989 saw another reversal – the dollar once again strengthening, contrary to general expectations. Throughout these swings, however, the dollar has not depreciated significantly against most developing nations' currencies, and this has encouraged a trend toward long-term investments in agricultural production in these nations, in expectation of expanded exports to US markets or in competition with US exports. In the short term, this has exacerbated domestic pressure in the United States for protectionism and export subsidies. In the long term, though, the continued instability of the dollar may eventually have the effect of reducing the incentives for such export-crop investments.

FINANCIAL INSTABILITY

Although there has been further integration of world financial markets, centring on the United States, a number of factors have led to great financial uncertainties. These include increased parity in interest levels and lower levels of inflation, exchange rate volatility and the persistence of high real interest rates. The stock market crash beginning in October 1987 was largely triggered by the chronic budget and trade deficits in the United States. Although stock markets have since appeared to recover, the financial implications of the recent surge in the dollar are unknown. If real interest rates do not fall sufficiently, then global recession may result. Developing countries with large external debts, and those exporting agricultural commodities, will suffer particularly.

TRADE WARS

There is a growing likelihood of serious trading frictions and retaliations between the United States, the EEC and Japan. An increase in general economic protectionism among the leading global economies could lead to declining world trade and economic activity. Again, indebted and agricultural-exporting developing countries are the most vulnerable.

The vulnerable economies

These stresses affect all developing economies, but in different ways. The risks are greatest for three particularly vulnerable groups: countries with predominantly agricultural exports, low-income food-deficit countries, and major agricultural exporters.

Countries with predominantly agricultural exports
Developing countries whose exports are predominantly agricultural commodities are extremely poor and highly indebted (Tables 3.1–3.4). In all countries with GNP per capita less than US$1,600 in 1985, over 50% of exports were agricultural commodities. In addition, most of these economies are solely dependent on one or two agricultural exports.

Low-income food-deficient countries
Although all 65 low-income food-deficit countries identified by the Food and Agricultural Organization of the United Nations (FAO) are potentially vulnerable in the current international climate,[9] those in the group that has failed to increase per capita food production are especially at risk (see Table 3.5). Growing food production in these countries is essential to meeting long-term food needs, as their extremely limited ability to purchase food imports is unlikely to improve in the near future. Moreover, data up to 1982 suggest that adequate growth in domestic cereal production in low-income food-deficit countries, and a healthy growth in export earnings, generally go together; and both are important determinants of the ability of a country to ensure food security. Rapid cereal production and steady growth in export earnings enable a country to raise levels of cereal consumption and achieve a degree of stability around the trend; in contrast, countries with sluggish increases in domestic cereal production tend to have inadequate export earnings, a low growth in cereal consumption and unstable cereal use.[10]

Major agricultural exporters
If protectionist pressure mounts and global markets for agricultural

Table 3.5: Low-income food-deficient countries with low food-production growth[a]

	Value added in agriculture (millions of 1980 dollars)		Cereal imports (thousands of metric tonnes)		Food aid in cereals (thousands of metric tonnes)		Average index of food production per capita (1979–81 = 100)
	1970	1985	1974	1985	1974/75	1984/85	1983–85
Africa							
Angola	–	–	149	377	0	78	102
Burundi	468	598	7	20	6	17	106
Cen. Afr. Rep.	256	333	7	17	1	12	105
Chad	416	–	37	134	20	163	106
Ethiopia	1,634	1,531	118	986	54	869	97
Guinea	–	805	63	140	49	47	102
Kenya	1,198	2,263	15	365	2	340	99
Lesotho	88	–	49	118	14	72	93
Malawi	258	426	17	23	–	5	105
Mauritania	200	222	115	240	48	135	94
Mozambique	–	477	62	426	34	366	98
Niger	1,466	1,070	155	247	73	218	96
Rwanda	295	614	3	24	19	36	106
Senegal	603	615	341	510	27	130	105
Somalia	589	911	42	344	111	248	102
Togo	238	325	6	79	11	23	103
Zambia	473	659	93	247	5	112	107
Near East							
Afghanistan	–	–	5	50	10	50	104
Sudan	1,754	1,511	125	1,082	46	812	103
Yemen, PDR	–	–	149	357	–	25	100
Far East							
Philippines	5,115	9,104	817	1,524	89	68	103
Sri Lanka	812	1,294	951	1,071	271	276	98
Latin America							
Bolivia	380	496	209	459	22	111	101
El Salvador	740	847	75	224	4	194	100
Haiti	–	–	83	227	25	101	104
Honduras	477	702	52	99	31	118	104

Notes

a   Low-income food-deficient countries as defined by FAO (1985).

– = figures not available.

Sources: FAO, Committee on Commodity Problems, 55th Session, *International Trade and World Food Security*, Rome, 21–25 October 1985; and World Bank, *World Development Report* (Washington, DC: World Bank, 1987).

commodities continue to be depressed, those developing countries with major shares in certain agricultural exports will suffer especially (see Table 3.6). Although the better-off exporters, such as Brazil and Argentina, may be able to absorb any resulting economic losses more easily, there are important implications for the structure and sustainability of their agricultural systems. The economic consequences will be even more severe, however, for those major exporters who are highly dependent on agricultural exports, have rising debt-servicing obligations and/or are increasingly dependent on food imports.

Examples of vulnerability

The increasing vulnerability of a developing country's agricultural system to external stresses and shocks can manifest itself in many complex ways. Several examples illustrate this point.

Indonesia
A recent study of the sustainable development of rainfed agriculture in the upper watersheds of Java in Indonesia indicates how long-term strategies for soil conservation and watershed management can be affected by changing agricultural export markets.[11]

Monocropping of cassava on erodible soils is generally discouraged by soil and water conservation projects in the Javan uplands because of the deleterious impact on soil structure. But concern over the decline in oil export earnings and mounting debt-servicing requirements has pushed the government of Indonesia (GOI) to expand all non-oil exports, including cassava. At present only 10% of Indonesia's cassava is exported. However, 97% of these exports are to the EEC, which has recently increased Indonesia's share of the total cassava import quota. As a result the GOI is concerned that a failure to meet its quota share will lead to a downward revision, even though in recent years cassava supply has been barely sufficient to meet domestic utilization. The government has thus promoted cassava exports very vigorously. The domestic price doubled in 1985 and again in 1987.

In response, farmers are switching from more sustainable and less erosive mixed-cropping and perennial crop-farming systems

Table 3.6: Country share of main developing-country agricultural exports, 1985[a]

(1) *Cocoa products*[b]

Ivory Coast	31.1	
Brazil	21.3	
Ghana	9.8	
Nigeria	8.0	
Cameroon	6.0	
All developing	97.8	Total world exports = $3,769m

(2) *Coffee*[c]

Brazil	20.7	
Colombia	15.5	
Ivory Coast	5.3	
Indonesia	4.9	
Mexico	4.7	
All developing	91.5	Total world exports = $11,467m

(3) *Tea*

India	24.3	
Sri Lanka	18.7	
China	13.4	
Kenya	9.8	
Indonesia	6.3	
All developing	86.4	Total world exports = $2,376m

(4) *Rice*

Thailand	26.7	
China	7.5	
Burma	2.6	
Indonesia	2.2	
India	1.9	
All developing	56.9	Total world exports = $3,104m

(5) *Sugar*

Cuba	53.1	
Brazil	4.1	
Thailand	2.6	
Mauritius	2.1	
Dominican Republic	2.0	
All developing	76.8	Total world exports = $8,923m

(6) *Bananas*

Honduras	17.3	
Costa Rica	13.4	
Ecuador	12.3	
Colombia	10.4	
Philippines	7.5	
All developing	93.8	Total world exports = $1,514m

(7) *Copra*

Papua New Guinea	25.9	
Malaysia	13.7	
Solomon Islands	12.2	
Vanuatu	9.9	
Singapore	9.2	
All developing	100.0	Total world exports = $131m

(8) *Groundnuts*

China	20.5	
Argentina	9.1	
India	5.0	
Vietnam	3.9	
Hong Kong	2.2	
All developing	52.7	Total world exports = $541m

(9) *Coconut oil*

Philippines	47.7	
Indonesia	15.7	
Singapore	6.6	
Malaysia	5.1	
Sri Lanka	4.8	
All developing	90.8	Total world exports = $728m

(10) *Groundnut oil*

Brazil	22.0	
Senegal	14.9	
China	14.2	
Argentina	7.8	
The Gambia	2.0	
All developing	67.1	Total world exports = $295m

(11) *Linseed oil*

Argentina	56.2	
Uruguay	1.4	
All developing	58.9	Total world exports = $146m

(12) *Palm oil*

Malaysia	60.2	
Singapore	21.0	
Indonesia	9.0	
Papua New Guinea	2.3	
Ivory Coast	1.4	
All developing	95.6	Total world exports = $2,641m

(13) *Jute and bast fibres*[d]

Bangladesh	76.7	
China	14.0	
All developing	97.4	Total world exports = $193m

(14) *Sisal*[e]

Brazil	42.6	
Kenya	26.5	
Mexico	10.3	
Tanzania	7.3	
Madagascar	4.4	
All developing	97.1	Total world exports = $68m

(15) *Rubber*

Malaysia	41.6
Indonesia	25.8
Thailand	18.0
Sri Lanka	3.4
Liberia	2.8
All developing	97.3 Total world exports = $2,783m

Notes
a   Defined as agricultural exports of which developing countries' share of world total is 50% or more. All figures are in terms of percentage of world exports unless otherwise indicated.
b   Includes cocoa beans, cocoa powder, cocoa butter and other products
c   Includes toasted and green coffee
d   Includes kenaf and allied fibres
e   Includes other hard fibres

Sources: FAO, *1986 FAO Trade Yearbook*, vol.40 (Rome: FAO, 1987); and FAO, *Commodity Review and Outlook 1986/87* (Rome: FAO, 1987).

---

to monocropping cassava. They are even removing terracing and other soil and water conservation structures to increase the area of cassava cultivation. On very steep slopes and highly erodible soils, the long-term impacts on land productivity may be severe. As domestic cassava prices eventually return to near world levels – they have in the past generally followed world market trends – the area planted to cassava in the Javan uplands should also decline. But, in the meantime, the on-site productivity costs and off-site erosion impacts of the recent price distortions may have already impaired the prospects for secure livelihoods for many upland farmers, and for the sustainable management of upper watersheds as a whole. In the long term, expanding cassava exports is neither an economically nor environmentally sustainable solution to Indonesia's growing debt problems.

Thailand
Certain major agricultural-exporting developing countries are facing increasing trade conflicts with industrialized competitors, particularly the United States. Thailand, for example, has long been one of the world's major rice exporters and is heavily

dependent on these earnings. But it is now facing strong competition from the United States which, in order to reduce rice surpluses produced by farm income-support programmes, has cut its rice export price in half over 1985–86. In an already depressed world grain market, Thailand now has to compete with the United States in the high-quality rice export markets for the EEC and Middle East, as well as with more traditional rivals, Pakistan and Burma, in the low-quality rice markets.

As a result, although Thailand's rice exports rose by nearly 330,000 tonnes in 1986, earnings declined by US$112 million. The average price per tonne dropped from US$215.85 to US$173.46. American rice exports to the EEC in 1986 increased by 33,000 tonnes, whereas Thailand's exports fell by 44,000 tonnes. Although world rice prices have since recovered to over US$200 a tonne, they are forecast to stay near this level for the rest of the century. This has led the World Bank and other multilateral lending agencies to stop financing public investments in irrigation and other infrastructure activities, which are crucial to expanded rice production in Thailand and other Asian countries. In the long term, this may eliminate the excess capacity of the global rice market, but in the meantime the United States still has 2 million tonnes of rice stocks and another 2 million tonnes of excess production capacity in the form of land presently diverted from production. The market for Thai rice will thus continue to be depressed by this excess capacity for some time.[12]

Ghana

The difficulty in reducing the vulnerability of the agricultural systems of low-income food-deficient economies to external economic stresses and shocks is illustrated by the case of Ghana. As with most poor Sub-Saharan African countries, whose growing indebtedness was attributed to poor economic management, Ghana was urged, in the early 1980s, to adopt structural adjustment reforms by the World Bank and IMF.

During the 1970s, agriculture (including forestry and fishing) contributed to over half of GDP, around 65% of average export earnings and 55% of labour force employment. Yet over this period, agricultural output declined at an annual rate of 0.3%, and real per capita income fell by over 30% in total. Major factors included a highly overvalued exchange rate and excessive export taxes on cocoa – by far the major export commodity. In addition, by 1982 producers were receiving less than 17% of the real price of cocoa in terms of 1962/63 values and half the level of producer prices in neighbouring Togo and Ivory Coast. Consequently, between 1960/65 and the early 1980s, Ghana's share of the global cocoa market fell from a peak of 36% to around 17%, with perhaps as much as 8–12% of cocoa output being illegally marketed through neighbouring countries.[13]

Beginning in 1983, Ghana attempted to reverse these trends through structural adjustment reforms which included a 90% exchange rate devaluation in real terms over 1982/83 to 1987, a 50% increase in the real price of cocoa over 1983–85 plus a further doubling by 1987, and a raising of the producer price of cocoa to over half the Ivory Coast level. The objectives were an immediate 25% increase in cocoa output by 1985/86 compared to two years earlier; and a long-term sustained output of 300,000 tonnes per annum.

Although cocoa output has increased substantially each year since 1983, reaching 230,000 tonnes in 1986, it still remains roughly 75% of 1975–80 levels. With so much of the agricultural recovery strategy relying on the revival of cocoa exports, the major constraint on production has been the depressed world market and the increased competition from new producers, such as Malaysia. Between 1973 and 1984 cocoa prices for Sub-Saharan exporters increased on average by only 0.3% per annum; and as a consequence the overall terms of trade for Ghana declined on average by 1.1% annually. They fell by as much as 50% between 1979/80 and 1983/84.[14] With cocoa accounting for close to 42% of total export earnings,[15] agriculture in Ghana – and indeed the whole economy – will remain vulnerable to fluctuations in the international cocoa market until there is a greater effort to diversify exports.

The unknown effect is the impact on the natural resource base. On the one hand, the general rise in rural incomes from the revival of small-holder cocoa, and thus all agricultural cultivation, may have reduced the motivation for over-extension of food production out of sheer poverty. On the other, the revival of cocoa's fortunes may have spurred some farmers to convert yet more environmentally-fragile marginal land to cocoa production.

Trade-offs

The above examples, as well as the preceding discussion, highlight an important question which must be considered in global policy-making for agricultural development. Namely, is there an inherent trade-off between the objective of global economic efficiency – which is being promoted by the West and Western-dominated multilateral institutions via pressure for free trade and structural reform in the developing countries – and national sustainability, and the equity between and within states? Unfortunately there is, at present, no clear answer to this question. The trade-offs undoubtedly occur under present policies but whether they are inevitable is another matter.

Although economic efficiency is the stated objective of current global economic and agricultural policies, this goal has yet to be realized. A recent World Bank report indicates that, if both developed and developing countries corrected the distortions caused by their present agricultural trade and pricing policies, the efficiency gains would be US$18.3 billion (1980 prices) for the developing countries, US$45.9 billion for industrial market economies and US$41.1 billion world-wide.[16] There is evidently a long way to go before global efficiency is attained. Moreover, it is also clear that present policies are hardly Pareto optimal let alone equitable between states. That is, it is possible, by ending these disturbing policies, to reallocate resources so as to make both industrialized and developing countries better off.

In addition, while it is apparent, as we have demonstrated above, that these reforms may threaten agricultural sustainability and have implications for the equitable distribution of income and wealth within the developing economies, our current knowledge and analysis of the potential trade-offs is insufficient to provide any idea of their magnitude or their inevitability. At this stage all we can say is that as these reforms progress over the coming years it is imperative that such analysis of trade-offs is pursued.

Finally, such an analysis cannot be completed on the basis of knowledge of international economic relations alone. Essential also is an understanding of how *national* policies influence the sustainability and equitability of agricultural development. It is to this issue that we turn in the next chapter.

Notes

1. World Bank, *World Development 1987* (Washington, DC: World Bank, 1987), tables 3, 32.
2. World Commission on Environment and Development, *Our Common Future* (Oxford: Oxford University Press, 1987), pp.13, 56, 142.
3. Leonardo A. Paulino, *Food in the Third World: Past trends and projections to 2000*, Research Report 52 (Washington, DC: International Food Policy Research Institute, 1986), pp.38–9.
4. World Bank, *Poverty and Hunger* (Washington, DC: World Bank, 1986). The report distinguishes between chronic and transitory food insecurity. Chronic food insecurity is a continuously inadequate diet caused by the inability to acquire food. It affects households that persistently lack the ability either to buy enough food or to produce their own. Transitory food insecurity is a temporary decline in a household's access to enough food. It results from instability in food prices, food production, or household incomes – and in its worse form produces famines.
5. ibid.
6. World Food Institute, *World Food Trade and US Agriculture, 1960–86*, 7th edn (Iowa: Iowa State University, 1987), pp.67–9.
7. Willem Buiter, *The Current Global Economic Situation, Outlook and Policy Options, With Special Emphasis on Fiscal Policy Issues* (London: Centre for Economic Policy Research, Discussion Paper Series No. 210, 1987).
8. World Bank, *World Development Report 1987* (Washington, DC: World Bank, 1987), pp.17–18.
9. See, for example, FAO, Committee on Commodity Problems, *International Trade and World Food Security*, 55th session (Rome, 21–25 October, 1985).
10. ibid.
11. World Bank, *Indonesia – Java Watersheds: Java Uplands and watershed management* (draft) (Washington, DC: World Bank, 1987). Both authors assisted in the preparation of this report.
12. Ammar Siamwalla, "Issues in Thai agricultural development" (draft) (Bangkok: Thailand Development Research Institute, 1987); and

American Embassy, Bangkok, *Thailand: Agricultural Situation Report – 1987* (Bangkok, 10 March 1987), pp.7–10. The latter report emphasizes (p.9) that other factors involved in the decline in the price of Thai rice exports in 1986 include a 43% increase in the export of low-quality rice due to government stocking policy; several large, high-quality markets for Thai rice such as Malaysia and Iran significantly reducing purchases; and discounts by Thai exporters.

13. World Bank, *Financing Adjustment with Growth in Sub-Saharan Africa, 1986-90* (Washington, DC: World Bank, 1986); and Wayo Seini, John Howell and Simon Commander, "Agricultural policy adjustment in Ghana", paper presented at the ODI Conference on The Design and Impact of Adjustment Programmes on Agriculture and Agricultural Institutions (London: Overseas Development Institute, 10–11 September 1987).

14. ibid.

15. World Bank, *Commodity Trade and Price Trends*, 1986 edn (Washington, DC: World Bank, 1986), table 9.

16. World Bank, *World Development Report, 1986* (Washington, DC: World Bank, 1986). East European non-market economies, which currently benefit from the low-priced global agricultural surpluses caused by current economic distortions in world agriculture, would lose US$23.1 billion.

4. National Policies

In the previous chapter we highlighted the international constraints with which developing countries must contend in devising more sustainable agricultural systems. Even though these constraints appear formidable, whether they actually become binding depends, to a large extent, on the national policies and strategies which developing countries adopt. In particular, if national agricultural strategies and targets fail to take account of the conditions required for sustainability and equitability – especially the need for proper resource management – the long-term prospects for agricultural development may be seriously undermined.

Agricultural strategies and targets

Strategies for agricultural development are difficult to typify for all the developing countries of the world. There are, none the less, certain key issues that are currently being highlighted in debates concerning appropriate agricultural targets and strategies for Third World development. These include:

- export versus food-crop production
- large versus small-scale farming
- the role of marginal versus more favourable agricultural lands
- the role of external assistance, and the role of the private versus the public sector.

In each case these raise, either explicitly or implicitly, problems of natural resource management and environmental degradation, on the one hand, and of social justice, distribution and participation on the other. In this chapter we begin by illustrating these cross-cutting problems by focusing on two of the key issues – export versus food-crop production, and marginal versus favourable land development.

Export versus food crop production

As part of the structural-adjustment policy reforms advocated by the International Monetary Fund (IMF), World Bank and other international lending agencies, many developing countries are being urged to reorientate their economies towards the production of tradable commodities, including agricultural exports. At the same time they are being urged to forego policies which promote self-reliance, for example increasing domestic food production to achieve self-sufficiency. Indebted African countries are being encouraged to specialize in export crops in which they enjoy a comparative advantage, as it is believed that their agricultural labour productivity is generally substantially higher in export than in food-crop production. It is also widely acknowledged that food-security needs can be met without a country having to be completely self-sufficient in food production. However, as we pointed out in the previous chapter, a high dependence on agricultural export commodities, and on food imports and/or aid, can leave a low or lower-middle income developing economy vulnerable to the external stresses and shocks imposed by the vagaries of international markets.

In addition, there are a number of arguments cautioning against overspecialization in agricultural export production, particularly in the African context. First, it may be difficult to achieve. Farmers tend to be risk averse, and may be unwilling to put their resources into export-crop production if their ability to produce adequate home food-supplies is in doubt. Second, a substantial portion of African and other developing country labour resources are already in food production. Thus, failure to raise significantly the productivity of these resources could mean leaving large numbers of people undernourished and in poverty during the long period required to shift to an alternative production and distribution system. Third, as the food-production resource base of developing countries varies considerably, particularly in Africa, the comparative advantage argument for export crop production may not apply to all places. Fourth, it is unlikely that any government, given reasonable prospects of success in domestic-food production, will find it politically acceptable to import the bulk of its basic food sustenance. Finally, specialization in export-crop production and reliance on increasing food imports may accelerate change in food preferences away from indigenous crops, such as millet, cassava and other root crops, to commodities readily available in international markets, such as rice, wheat and maize. If such supplies fail, for any reason, it may prove difficult to revert to indigenous food crops. Over a period of time much of the knowledge associated with indigenous food production may be lost or, at the least, the technologies will have stagnated.

On these grounds it is often argued that the best strategy is a mixed one, emphasizing continued promotion of food-crop production coupled with selective specialization in export crops to boost foreign exchange.[1]

However, there is an additional dimension to the controversy. A frequent criticism of policies to promote export-crop production in developing countries is that export-orientated agricultural development is less environmentally sustainable than food production for domestic consumption. This is an important issue but we believe the argument is too simplistic. In our view the main obstacle to sustainable agricultural development is the failure of any economic policy, whether promoting food crops or exports, to address adequately problems of natural resource management. Policies to achieve food self-sufficiency may therefore be neither inherently more nor inherently less environmentally sustainable than export-orientated agricultural development.[2]

One major difficulty in analysing the sustainability of cash-versus food-crop 'production is that distinctions between cash crops and food crops are not clear cut. Often the terms "cash crops" and "export crops" are used synonymously. Strictly speaking, however, a cash crop may be sold at home or abroad and may be either a food or non-food commodity, whereas an export crop is a cash crop which ultimately is exported from the country producing it. The major non-food cash crops which are exported are cocoa, coffee, fibre crops, rubber, tea and tobacco. In contrast, the term "food crop" usually refers to domestic production of basic staples (cereals, pulses, roots and tubers). Although these are the principal subsistence crops, they are also often marketed.[3] For example, in Asia sizeable proportions of rice and wheat, which are basic food staples, are sold for cash. Rice is a major export crop for Burma, China, Pakistan and Thailand.

Moreover, aggregate evidence suggests that expansion of cash cropping for export in most developing countries is not necessarily at the expense of staple food production. In general, countries tend to manage sufficient growth in both cash-crop and staple-food production or fail to achieve either (see Table 4.1 and discussion in previous chapter). For example, in Sub-Saharan Africa, constant or declining per capita food production has been associated with constant or declining shares of land allocated to cash crops. As agricultural export earnings stagnated or declined in most of Sub-Saharan Africa, countries in the region were unable to import sufficient agricultural inputs, spare parts, or raw materials and consequently also experienced falls in per capita food production. Over 1968–82, the majority of countries with positive growth in per capita production of basic staples have simultaneously expanded their area devoted to cash crops.[4]

The crucial questions are: where is the expansion occurring? And with what specific crops? The amount of land growing both export and food crops in developing market economies has increased in the last ten years due to the bringing into production of "new" land, such as areas under forest or previously considered marginal (see Table 4.2). In some instances, the expansion of cash cropping for export – such as in the southern Volta region of Ghana and the Cauca Valley of Colombia – may take the most fertile land, pushing food production and subsistence farming on to marginal lands. In other regions, government policies deliberately encourage the production of food crops in marginal areas, often without simultaneously encouraging proper management techniques and agricultural practices which can reduce environmental and soil-erosion problems. In Haiti pricing policies have encouraged the

Table 4.1: Changes in production of cash crops compared with changes in production of basic food staples, by region, 1968–82

Growth in share of cash-crop area in total land use	*Growth in per capita food production per year (number of countries)*			
	Less than −1%	*+/−1%*	*More than 1%*	*Total*
Asia and Pacific				
Less than −1%	1	2	1	4
±1%	3	3	5	11
More than +1%	1	4	3	8
Total	5	9	9	23
Africa				
Less than −1%	5[a]	6[b]	1[c]	12
±1%	7[d]	6[e]	1[f]	14
More than +1%	4[g]	3[h]	2[i]	9
Total	16	15	4	35
Latin America and Caribbean				
Less than −1%	1	0	1	2
±1%	3	4	6	13
More than +1%	1	2	2	5
Total	5	6	9	20
All Countries				
Less than 1%	8	8	2	18
1%	13	12	13	38
More than 1%	6	9	7	22
Total	**26**	**30**	**22**	**78**

Notes: The rates of change are annual changes in estimated trend lines. Grains, pulses, roots and tubers (in grain equivalents) are included. Totals rounded.
a   Chad, Mali, Mozambique, Togo and Uganda
b   Benin, Central African Republic, Congo, Nigeria, Sierra Leone and Upper Volta
c   Niger
d   Angola, Ghana, Guinea, Kenya, Malawi, Morocco and Somalia
e   Burundi, Ethiopia, the Ivory Coast, Liberia, Senegal and Zaire
f   Tanzania
g   Egypt, Madagascar, Mauritania and Zambia
h   Cameroon, Rwanda and Zimbabwe
i   Sudan and Tunisia

Source: Joachim von Braun and Eileen Kennedy, *Commercialization of Subsistence Agriculture: Income and nutritional effects in developing countries*, Working Papers on Commercialization of Agriculture and Nutrition, no.1 (Washington, DC: International Food Policy Research Institute, 1986).

growing of maize and sorghum in hilly areas at the expense of coffee and other tree crops, and have increased soil runoff and erosion.[5] Similarly, throughout the Third World, the planned extension of maize, sorghum and millet into dryland areas has tended to exacerbate problems of soil erosion and exhaustion. Tables 4.3 and 4.4 indicate that in both tropical and sub-tropical (e.g. Javan upland and West African) conditions, land under these and other annual food crops may be more susceptible to erosion than under other forms of vegetation cover.

Table 4.2: Harvested areas under basic food and export crops (million hectares)

	All developing countries		Africa	
	1974–76 *(average)*	*1984*	*1974–76* *(average)*	*1984*
Food				
Cereals	301.9	322.2	69.7	70.9
Roots and tubers	20.7	23.1	11.2	13.0
Pulses	46.9	51.3	11.7	12.6
Total	**369.4**	**396.5**	**92.6**	**96.4**
Export				
Cotton	20.2	20.7	4.0	3.9
Coffee	8.6	10.1	3.3	3.3
Cocoa	4.4	4.9	3.2	3.3
Tea	1.0	1.3	0.1	0.2
Tobacco	2.3	2.2	0.3	0.3
Sugar	11.4	15.0	0.5	0.6
Palm Oil	3.9	4.8	0.7 (est)	0.9 (est)
Rubber	5.6	6.5	0.2 (est)	0.2 (est)
Total	**57.4**	**65.5**	**12.3**	**12.7**

Source: UN Food and Agriculture Organization Production Yearbook 1984 (Rome: FAO, 1985), plus additional FAO figures.

In summary, agricultural strategies which do not take into account the possible environmental impacts and displacement effects of increased production may lead to a less than optimal allocation of land use in both the short term and long term.

Table 4.3: Vegetal cover factors (C) for erosion in West African conditions

	C
Bare soil	1.0
Dense forest or culture with a thick straw mulch	0.001
Savannah and grassland, ungrazed	0.01
Forage and cover crops – late-planted or with slow development:	
First year	0.3–0.8
Second year	0.1
Cover crops with rapid development	0.1
Maize, sorghum, millet	0.3–0.9
Rice (intensive culture, second cycle)	0.1–0.2
Cotton, tobacco (second cycle)	0.5
Groundnuts	0.4–0.8
Cassava (first year) and yams	0.2–0.8
Palms, coffee, cocoa, with cover crops	0.1–0.3

Notes: C. representative annual value. The C value indicates the rate of erosion under different cropping patterns and cover, relative to bare soil. In general, the better the protection of the soil surface, the lower is the rate of erosion.

Source: Robert Repetto, *Economic Policy Reform for Natural Resource Conservation* (Washington, DC: World Resources Institute, 1986).

Moreover this is true whether the aim of the strategies is to promote export-crop production or achieve food self-sufficiency.

Favoured versus marginal lands

For marginal lands, farming systems need to be chosen to suit both the given agro-ecological conditions and the economies of local farming households. Too often, though, production-led policies for both food and export crops are designed without sufficient knowledge of these conditions and of their implications, particularly for sustainable agricultural development.

A first step in addressing this issue is to make a clear and explicit distinction between "green revolution" agriculture on more favoured agricultural lands, i.e. areas which are generally fertile, irrigated or otherwise well-watered, uniform and flat; and "low-resource" or "resource-poor" agriculture on more marginal agricultural lands, i.e. areas which are generally less fertile, rainfed,

---

Table 4.4: Estimated soil loss for different land uses (Java – middle volcanic agroecosystem)

Type of land use (volcanic soil)	*Measured soil loss (Tonnes/ha/growing season)*
Bare soil, terraced	20
Cabbage, terraced	15
Maize, non-terraced; 3-degree slope	7
Grass, terraced	0.3
Coffee, non-terraced; 3-degree slope	0.2

Source: Brian Carson and Wani Hadi Utomo, *Erosion and Sedimentation Process in Java* (Malang, Indonesia: KEPAS and the Ford Foundation, 1986).

---

diverse and undulating. Marginal lands in the Third World, which are typical of most of Sub-Saharan Africa but also the semi-arid and arid lands, uplands, swamplands and converted forest-lands of Asia and Latin America, are characterized not only by lower quality and productivity but also by their greater insecurity. This is especially true of their microclimatic, agro-ecological and soil conditions. Moreover, changes in marginal farming systems – such as the introduction of productivity-increasing technologies and crop specialization – which are not adequately adapted to these conditions, may actually impose additional stresses that make the systems even more vulnerable. This often means that, irrespective of whether the general productivity trend is upward or downward, its variability and the frequency and seriousness of crisis situations may increase and threaten overall sustainability of production.[6]

The sustainability of resource-poor agriculture on marginal lands is crucially linked to the responses of poor rural people to population growth, migration and "core invasions and pressures". The latter can be defined as "extensions into rural areas of the power, ownership and exploitation of central, urban institutions and individuals which include the richer world of the North, governments of the South, commercial interests, and professionals who are variously wealthy, urban and powerful".[7] Although rapid population growth and uneven distribution in some areas of the Third World – particularly on marginal lands – undoubtedly complicates natural resource management, it is also true that:

> Population pressures on resources usually reflect an extremely skewed distribution of resources. When farmers encroach on tropical forests or cultivate erodible hillsides, population pressure is blamed, but the pressure typically stems from the concentration of land in large holdings.[8]

As depicted in Figure 4.1, the responses of the rural poor in marginal areas to such external and internal stresses can have important development consequences:

> In such areas, as populations grow and common property resources are appropriated, agriculture becomes more intensive, and for a time

Figure 4.1: Rural migration and resource exploitation: some general tendencies

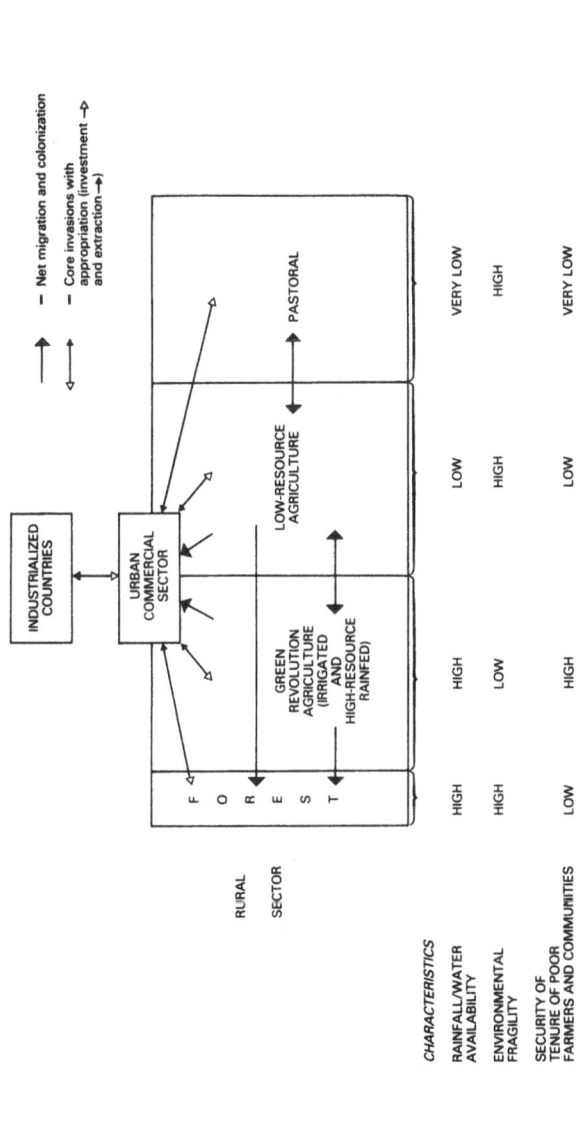

Source: Robert Chambers, "Sustainable livelihoods, environment and development: putting poor rural people first", Discussion Paper 240 (Brighton: Institute of Development Studies, University of Sussex, 1987).

at least, less sustainable as fallows shorten and/or livestock become more numerous. Core invasions and pressures, appropriations and exclusions by governments and by the urban and rural rich, declining biological productivity, and rising human populations may drive many of the poorer people to migrate. This they do either seasonally or permanently, some to cities and towns, some to areas of green revolution agriculture, and some to forests, savannas, steep slopes, flood-prone flatlands and other vulnerable or marginal areas.[9]

The failure to provide adequate, secure and sustainable liveli-hoods for the rural poor in marginal areas therefore exacerbates existing problems of population pressure, rural–urban migration, degradation of fragile environments and, even, political instabil-ity. The answer, though, is not to abandon marginal lands, but "to see how more people can gain such livelihoods where they are already, without having to migrate to towns or other rural areas where they so often suffer and aggravate already bad conditions for others."[10]

One of the main biases in agricultural development strategy is the assumption that resource-poor agriculture in marginal areas has limited production potential. Yet evidence suggests this assumption is false. Although the productivity of marginal lands may not reach the high yields of more favoured lands, experi-ence shows that a combination of appropriate farming-systems techniques, research and extension, inputs, economic incentives, infrastructure and, above all, participation and commitment by the beneficiaries, can lead to successful projects under the most difficult agricultural conditions. One of many examples is a World Neighbours project in Honduras, discussed further in Chapter 5.[11] At a cost of US$13 per person, the Guinope Integrated Development Program has transformed a previously unsustainable small-holder agroecosystem, through appropriate agricultural technology, training and erosion control – includ-ing intercropping of "green manure" crops with the traditional corn or sorghum – into a surplus-producing system with yield increases of over 300% and a marketable surplus of vegetables.

Even in drought-prone Sub-Saharan Africa, there are numerous successes in improving agricultural sustainability in resource-poor systems – ranging from the large-scale Kenyan Soil and Water Conservation Programme to the Yatenga Water Harvesting Project in Burkina Faso.[12] Replicating these successes on a larger scale and in other areas could have a significant impact on the production potential of marginal lands.

In addition, the current contribution of production on marginal lands may already be important to agriculture, even in those countries fortunate to have large areas of more favourable lands. For example, rice production in Indonesia, which accounts for 69% of the total food-crop area harvested, already occupies the most fertile lowland areas on the islands of Java, Bali, Southern Sulawesi and Southern Sumatra. On these lands there is little room for expanding irrigated rice production or increasing yields, suggesting that agricultural resources there are already being exploited at or near their full potential production levels.

In contrast, agricultural production on Indonesia's more marginal lands is characterized by low yields. The causes are unsuitable cropping systems, land-management techniques, input packages and, above all, research and extension advice that are inappropriate for the more diversified and fragile agro-ecological conditions found on these lands. Nevertheless, dryland – mainly upland – food production accounts for nearly two-thirds or more of maize, cassava, sweet potato and peanut production and around 40% of soybean production on Java. The total dryland area planted to paddy and secondary crops on Java amounts to about one-fifth of the total harvested food-production area in Indonesia. Thus food production on the marginal drylands of Java alone may contribute to over 3% of gross domestic product (GDP) and about 15% of agricultural GDP. If the currently low dryland yields were to be substantially increased by introducing appropriate techniques, inputs and advice, food production on these lands could rise by 25%. In particular, on highly erodible soils (e.g. limestone clays) and on slopes greater than 50%, switching out of annual food cropping altogether into perennial tree crops and livestock-based systems would significantly increase the economic potential of severely degraded uplands.[13]

Appropriate technology for marginal lands is not, however, enough by itself. Whether the full economic potential of such lands is realized will depend equally on designing appropriate pricing policies.

Pricing and macro-economic policies

Current agricultural strategies in developing countries, whether aimed at export promotion or food production, tend to be narrowly focused on maximizing short-term gains with very little regard to proper resource management. They are often translated into very singular production goals and targets. Issues of sustainability – whether patterns of resource use can sustain increased production, or whether investment programmes and incentive schemes are contributing to such problems as soil erosion, water scarcity and deforestation – are not generally given high priority. Nor, as agro-processing and domestic production of agricultural inputs are developed, is there usually any concern for the additional problems of competition over scarce water supplies, pollution and the handling and disposal of toxic wastes.

As developing economies are generally characterized by active government intervention in markets, such strategies are reinforced by, if not enacted through, pricing and macro-economic policies. Of particular importance are:

- macro-economic policies, such as trade, exchange rate, fiscal and monetary policies, which have a significant impact on agriculture
- agricultural input and output pricing policies and interventions, such as export and import duties, subsidies, producer margins and government monopolies
- agricultural stabilization policies, including the use of consumer subsidies, price stabilization, marketing boards and manipulation of stocks.

These policies can influence the relative prices of agricultural goods and also change the prices of all agricultural products relative to those of non-agricultural goods and services. They are, therefore, powerful determinants of the sustainability of agricultural development in Third World economies. If policy strategies in these countries are to incorporate natural-resource management concerns, then it is also necessary to develop complementary changes in macro-economic, pricing and regulatory interventions.

This is important for two reasons. First, these policies have a direct and far-reaching impact on producer and consumer economic activity, which in turn affects the incomes, resource allocation and investment decisions of individual farmers. For example, if farm profitability is reduced:

> the returns on investments in farmland development or conservation are also depressed, which reduces both the farmers' ability and their incentive to invest in levelling, terracing, drainage, irrigation, and other land improvements. The resulting loss of land productivity through erosion, salinisation, and depletion of nutrients compounds the problem of rural poverty, even in the short run.[14]

Second, the benefits of agricultural investments, including investment programmes supporting environmental and natural-resource management objectives, are sensitive to changes in overall economic and agricultural policies. For example, soil and water conservation projects which emphasize the use of agroforestry systems based on high-valued perennial crops – such as coffee, cocoa, rubber, bananas, tea and spices – in order to provide root structure and canopy cover on erodible upland soils, have less chance of succeeding if the returns to producers are reduced by export taxes, monopoly marketing practices or overvalued exchange rates.

Exchange rates
Of particular concern are macro-economic policies which are biased against agriculture. Overvalued exchange rates and high levels of protection to non-agricultural sectors effectively place a tax on farming. The prices of industrial import substitutes, farm inputs and nontraded goods are increased relative to the prices of agricultural import substitutes and exports. The internal terms of trade for agriculture deteriorate and agricultural prices become chronically depressed, so undermining efforts to design policies to encourage more sustainable agricultural development.

These effects are especially apparent in Sub-Saharan Africa.

Table 4.5: Index of nominal and real protection coefficients for cereals and export crops in selected African countries, 1972–83

| | *Cereals* | | | | *Export crops* | | | |
| | *1972-83* | | *1981-3* | | *1972-83* | | *1982-3* | |
Country	*NI*	*RI*	*NI*	*RI*	*NI*	*RI*	*NI*	*RI*
Cameroon	129	90	140	108	83	61	95	75
Ivory Coast	140	98	119	87	92	66	99	71
Ethiopia	73	55	73	49	88	71	101	66
Kenya	115	94	115	98	101	83	98	84
Malawi	85	79	106	100	102	94	106	97
Mali	128	79	177	122	101	83	98	70
Niger	170	119	225	166	82	59	113	84
Nigeria	126	66	160	66	108	60	149	63
Senegal	109	79	104	89	83	60	75	64
Sierra Leone	104	95	184	143	101	93	92	68
Sudan	174	119	229	164	90	63	105	75
Tanzania	127	88	188	95	86	62	103	52
Zambia	107	93	146	125	97	84	93	80
All Sub-Saharan Africa	**122**	**89**	**151**	**109**	**93**	**71**	**102**	**73**

Note: The nominal index measures the change in the nominal protection coefficient with border prices converted into local currency at official exchange rates. The real index measures the change in the nominal protection coefficient with border prices converted into local currency at real exchange rates. Data for Ghana are not available.

NI = nominal index
RI = real index

Source: World Bank, *World Development Report 1986* (Washington, DC: World Bank, 1986).

Between 1969–71 and 1981–83, for all Sub-Saharan African countries, real exchange rates appreciated by 31%. Exchange rate overvaluations were particularly large in Ghana, Nigeria, Tanzania and Uganda. Over the same period, those countries whose currencies showed an annual average rate of appreciation experienced a lower growth rate in agriculture (1.5% growth per year, or 1.1% excluding Botswana) than those whose currencies depreciated (2.6% growth per year). Moreover, although between 1969–71 and 1981–83 incentives for cereal production in these countries – calculated in terms of nominal protection coefficients using official exchange rates – increased by 51%, when real appreciations in currencies are taken into account, the actual increase in incentives was only 9%. For export crops, incentives increased nominally by about 2% but due to currency appreciation actually declined in real terms by 27% (see Table 4.5).[15]

Food prices
Over the long term, higher food prices could lead to higher rural real wages generally and improve the efficiency of resource-allocation and use, thereby generating economic growth and increased employment. None the less, relying on this as a means of inducing economic behaviour more conducive to sustainable agriculture may have uncertain and unintended impacts, particularly on equitability. For example, it is often believed that increasing domestic food prices provides greater incentives to farmers to increase the supply and sustainability of food production, by providing them with a secure income to invest in farming-system improvements which reduce environmental degradation. However, it may have the unintended result of reduced real incomes and creation of severe hardships for the poor, at least in the short term.

As indicated in Table 4.6, both urban and rural poor are more responsive to food price changes than higher-income groups. In particular, landless rural labourers suffer from higher food prices, as they are increasingly paid with cash rather than goods, and their wages change more slowly than do prices. Many of the rural poor do not derive a large share of their income from either wage labour in food production or from the sale of food, and a large proportion are *net consumers* of food. Often those who do have access to land produce food largely for themselves and have only a small marketed surplus.

Thus the prospect of uncertain gains may be insufficient compensation for the poor who, in the short term, are adversely affected. Moreover, market imperfections and government policy which reduce producers' share of the final price may weaken even the long-term impact of higher food prices in terms of higher farm incomes and increased farm output.

Although remunerative food prices as supply incentives are essential for sustainable agricultural development, raising food prices alone without complementary structural reforms may actually exacerbate poverty. Such reforms include improvements in the efficiency of food marketing, increasing producers' price margins and the introduction of cost-reducing technological improvements which are appropriate to local agroecological

Table 4.6: Price elasticities of demand for rice among low-income and high-income groups, selected countries

Country	Low income Percentile	Low income Price elasticity	High income Percentile	High income Price elasticity
Bangladesh (rural)	10	−1.30	90	−0.83
Brazil	15	−4.31	90	−1.15
Colombia (Cali)	1	−0.43	93	−1.19
India (rural)	3	−1.39	96	−0.39
India (urban)	1	−1.23	92	−0.21
Indonesia	8	−1.92	55	−0.72
Philippines	12	−0.73	87	−0.40
Sierra Leone (rural)	16	−2.16	84	−0.45
Thailand	12	−0.74	87	−0.46

Source: Per Pinstrup-Andersen, "Food prices and the poor in developing countries", in J.P. Gittingen, J. Leslie and C. Hoisington (eds), *Food Policy: Integrating supply, distribution and consumption* (Baltimore: Johns Hopkins University Press, 1987).

conditions and resource endowments even in rural areas, without benefiting producers significantly. But because such reforms may take time to expand production and hold down prices, selectively targeted food subsidies must be used to ease poverty temporarily. In the long run, the hope is that expanded food production will result in less expensive food, as well as a reduction in any food imports. Furthermore, the economic gains from the temporarily higher food prices in conjunction with long-term structural reforms should exceed the cost of full compensation to the poor in the short term.[16]

Input subsidies
Input subsidies aimed at increased food and cash-crop production may also have important and unanticipated impacts, in this case on efficiency and sustainability. In Indonesia, for instance, subsidies for fertilizers have reached 68% of world prices. As a result, consumption of fertilizer increased by 77% (12.3% per year) over 1980–85. The current rate of consumption, 75 kilogrammes per hectare (kg/ha) of arable land, is much higher than in other Asian countries (e.g. 32 kg in the Philippines and 24 kg in Thailand), and is encouraging inappropriate application and wastage. Similarly, pesticide subsidies of 40% and irrigation subsidies of 87% in Indonesia are encouraging wasteful use of these inputs.[17]

In addition to imposing a financial burden on Third World governments, inappropriate input subsidies for fertilizer, pesticides and irrigation can impose considerable external costs in terms of agricultural pollution and resource depletion. Some of these can be considered user-costs; the farmer may lose future agricultural productivity because of pesticide resistance, or through misallocation of input investment or inappropriate use, or because of future scarcity of resources such as water. In Indonesia the total losses in irrigated rice production from the 1986/87 outbreak of brown planthopper attack is an estimated US$390 million.[18] But inappropriate use of agricultural inputs also produces a wide range of negative externalities. These include damage to human health, fisheries and biological diversity through pesticide misuse; problems of groundwater contamination and eutrophication of surface water from fertilizer run-off; and the diversion of scarce water supplies to irrigation from other valuable uses (e.g. industrial purposes, domestic use and fish ponds). The environmental implications of agricultural input subsidies are rarely considered in the design of agricultural policies, yet the user and externality costs of these impacts are often very high. In all developing countries, the environmental dimension of major changes in economic incentives needs to be more thoroughly analysed, if policies for more sustainable agricultural development are to have a realistic chance of success.

Pricing policy and sustainability: Indonesia

Indonesia provides a good case for such an analysis. In particular, it clearly illustrates that governments can dramatically influence the incentives for sustainable agricultural development. Policies on commodity prices, farmer incentives and input subsidies all have significant implications for erosion, pollution and the use of scarce resources.[19]

Commodity prices
Agricultural markets in Indonesia are highly complex, and although government management is pervasive, the degree of intervention varies significantly from market to market for the various crops cultivated. Some crops are protected while others are not. The market for rice, for instance, is highly regulated, with the government of Indonesia's (GOI) procurement agency, BULOG, maintaining floor and ceiling prices through its accumulation and control of inventory stocks and imports. BULOG has also been active in the markets for sugar, corn, soya beans and wheat, although mainly to restrict imports. In addition, extremely high and effective protection rates exist for fruits, vegetables and dairy products as a stimulus to domestic production, which for the most part is not traded internationally. The rate has been as high as 200%.[20] Cassava, too, has been supported. Prices doubled in 1985 and again in 1987, largely reflecting the GOI's targeting objectives of overcoming domestic shortages and procuring sufficient supplies to meet the EEC export quota (see also Chapter 13).[21]

In contrast, there has traditionally been little government intervention in the markets for non-grain staples (apart from cassava), such as groundnuts and minor legumes (mungbeans, pigeon peas and so on), which are mostly nontradables.

Examination of the ratio of domestic producer prices to border prices (the nominal protection rate – NPR) suggests that despite the varying degrees of market intervention, prices for rice, corn and cassava have not been significantly distorted. But the mainly positive NPRs for soya beans and sugar between 1972 and 1985 indicate that import controls have lifted domestic prices well above world levels. For export crops the long-term decline in world commodity prices has significantly eroded the nominal and real incentives to domestic producers, but recent devaluations have somewhat restored Indonesia's competitiveness.

The overall effect of these interventions has been to reinforce the profitability of horticultural crops and, to a lesser extent, soya beans and livestock products. Protective pricing together with rigid import controls and stringent area-targeting have also resulted in expanded small-holder sugar production on Java. And there have been steady increases in rice production, although less a function of producer prices, which have been declining in real terms, than of input subsidies. This, however, has had the effect of depressing prices for the less desirable staple substitutes produced mainly on rainfed lands, such as corn and root crops. They are strong substitutes for rice, especially among the rural poor.

Pricing and the environment
What have been the environmental implications of this agricultural pricing structure, particularly in the uplands of Java? The most notable effect arises from the dramatic increase in terms of trade for horticulture and livestock products. These appear, over the long term, to be encouraging upland agricultural production to move from less profitable cultivation of relatively inelastic, basic starchy staples to more profitable, income-elastic commodities such as fruits, vegetables, milk and meat. This may constitute an important incentive for upland farmers to invest in soil conservation

measures and improved land-management techniques, although increased profitability alone may not be sufficient.[22] However, the increased profitability of vegetable crops also means that farmers are encouraged to cultivate them on steeply sloped volcanic soils, where water run-off and hence soil erosion are enhanced.[23] Furthermore, as the average returns increase to these and other highly commercialized and input-intensive crops, such as sugar cane, then share tenancy and absentee ownership become more common. If these tenancy arrangements are insecure or if the objective of absentee owners is short-term profit maximization or land speculation, then incentives for long-term investments in improved land management may be greatly reduced.

Finally, the recent and rapid rise of cassava prices is worrying, as some upland farmers are switching back from more protective farming systems, based on livestock rearing, agroforestry and multi-cropping, to growing cassava alone on highly erosive soils (see Chapter 3).

Farmer investments
To what extent are farmers making long-term investments? Improvements in terms of trade may not be directly benefiting farmers who need to make these investments. Although in the last few years the relative competitiveness of agricultural exports has improved due to devaluations, the considerable market power of exporter associations, licensed exporters and approved traders and other marketing intermediaries ensures that upland farmers are receiving relatively few of the benefits.[24] In general, farmers growing crops on marginal lands tend to have lower producer margins than farmers growing crops on the irrigated lowlands. For example, farmers receive 80–85% of the retail price for rice, 70–75% of the retail price for soya beans and only 60–65% of the final price for corn, which is predominantly a dryland crop.[25] Farmers on marginal lands are less likely to engage in marketing activities and more prone to price discrimination by marketing intermediaries. In the Citanduy River Basin, West Java, only 10–20% of clove and peanut farmers perform marketing activities, such as drying or transporting the commodities to sub-district sellers.[26]

In addition, while pricing policies can encourage sustainable agricultural practices they are rarely, by themselves, sufficient to ensure that new appropriate farming systems for marginal lands are adopted. For instance, on steep uplands livestock- and agro-forestry-based systems are likely to be more sustainable than the cultivation of annual crops, while on acidic swamplands coconut-based systems may be more appropriate than irrigated rice. Yet if diversified small-holder production systems such as these are to be viable on marginal lands, improvements are needed in the quality and marketing of small-holder production, particularly of potentially tradable crops and import substitutes. As an example, improved drying of coconut would increase the value of copra by at least Rp 25/kg, but for small-holders this requires knowledge of better techniques (such as using the coconut shell for drying and not using the coconut husk for fuel) and collective investment, such as farmers' groups sharing the costs of more efficient drying kilns.[27]

Input subsidies
What are the effects of such subsidies on sustainable practices? In Indonesia, input subsidies total about US$725 million in 1985. The current effective subsidy for fertilizers to farmers is about 38% of the farmgate price (68% of the world price); for pesticides more than 40%; for irrigation as much as 87%; and for credit an implicit rate of 8%.

This policy of heavily subsidizing agricultural inputs was one of the hallmarks of the rice self-sufficiency strategy of the 1960s and 1970s, and thus the bulk of the subsidies has benefited the lowland, irrigated, mainly rice-producing areas on Java, South Sumatra, South Sulawesi and Bali. The effects have been dramatic; the area of higher yielding varieties (HYVs) has expanded from 0.8 to 6.8 million ha, and on Java the average area planted with HYVs has reached 94%. The irrigated area increased from 3.7 to 4.9 million ha. Distribution of subsidized fertilizers rose from 0.2 to 4.1 million tonnes, and of subsidized pesticides from 1,080 to 14,210 tonnes.[28]

Now, with a new emphasis on agricultural diversification, these subsidies are increasingly being used to stimulate production of other crops – notably sugar, cassava, maize, palm oil and soya beans. Assuming no change in input policy, the total cost of the subsidies is anticipated to increase, as they are gradually extended to agricultural cultivation on marginal lands. For example, rainfed crops on Java, with the exception of high-valued vegetables, fruits and estate crops, still tend to use relatively fewer subsidized inputs than irrigated rice and sugar, but use relatively more organic fertilizers. This will change, but not necessarily for the better.

Although the yields and net returns of intensive irrigated rice on Java are substantially higher than for rainfed crops, this does not imply greater efficiency in use of inputs. For instance, despite the larger applications of chemical fertilizer and pesticides on intensive irrigated paddy, their use on non-intensive irrigated paddy and on the predominantly rainfed staple crops, apart from maize, appears to incur lower per-unit costs. This suggests that subsidies are encouraging overuse of these inputs in intensive rice cropping. Similarly, per-unit irrigation costs for rice are strikingly low, given that irrigation accounts for 91% of the water use on Java.[29] Efficiency of input use is thus likely to decline and be accompanied by a greater and more widespread environmental impact, particularly from fertilizers and pesticides.

Fertilizers
Overuse of fertilizers is already a substantial problem in lowland irrigated areas. In some areas of Indonesia, applications of urea can reach 200–250 kg. Since fertilizer comprises less than 10% of the production cost of rice and the largest production response is achieved at relatively low levels of application, the current high rice–fertilizer price ratio of 1.5–2 will continue to encourage inappropriate application and waste, with little stimulation to rice output.[30] Moreover, providing subsidized fertilizers to cultivators on marginal lands may be counter-productive, in that farmers will apply relatively cheap fertilizers to increase yields, rather than consider more expensive but environmentally sound methods such as green manuring, mulching and using compost to maintain soil fertility. For example, in Ngadas, East Java, farmers are presently using over 1,000 kg of subsidized chemical fertilizers per hectare to produce two 10-tonne potato

crops. Yields are declining and, as experiments have shown, are less than one half of what could be attained with improved soil management and green manuring techniques. Recently, the farmers have come to realize that increased fertilizer use was not offsetting yield reductions and have begun to use more organic fertilizer.[31]

Pesticides
The government has recently banned the use of 57 pesticides and is undertaking an integrated pest-management training programme with the World Bank and the Food and Agriculture Organization (FAO). Nevertheless, the current subsidy levels will most likely continue to encourage inappropriate and excessive use. In fact, the pesticide ban was a belated response to the latest plague of rice brown planthopper, which was associated with misapplication of pesticides that have wiped out natural enemies of pests. Pesticide subsidies tend to discourage traditional methods of eradicating pests and make integrated and biological pest-control methods relatively less attractive to farmers. Subsidized pesticides encourage farmers to treat fields preventively even before an economically damaging insect population is present, causing natural enemies to be killed and releasing pests (e.g. brown planthopper) from natural control. Even rice varieties normally resistant to brown planthoppers, such as IR–36, have been known to be "hopperburned" (severely damaged from brown planthopper feeding) when treated too often with insecticides. For example, in Northern Sumatra, the population density of brown planthopper (between 0.5 and 40 per plant) rose directly as the number of reported insecticide applications increased; in five areas experiencing hopperburn farmers were treating fields six to twenty times in 4–8 weeks without any success.[32] Recently there have been attempts to reduce pesticide subsidies. But while fiscal outlays for the subsidies have been reduced, preliminary indications suggest that the costs of these subsidies are being shifted from the official budget to the operations of parastatal producers, who are financing the cost burden through additional borrowing.

Irrigation

The high level of subsidy for irrigation – US$401 million spread over approximately 4 million hectares – is also causing problems of overuse. With total operation and maintenance spending being reduced by budget cuts, the failure to recover any significant amount of irrigation costs is also jeopardizing the supply network. In the long run, failure to maintain the irrigation network will translate into losses of agricultural productivity, which will be exacerbated by any water scarcity problems caused by overuse. Allocation of scarce water supplies will become a pressing problem on Java in the near future, as municipal and industrial uses continue to expand.

Credit

Credit is of crucial importance in furthering adoption of improved soil-conservation and land-management techniques on marginal lands. For example, investments in bench terracing require a medium-term loan for at least two years and short-term loans for succeeding years. Agroforestry requires long-term loans for at least seven years. Different rates and terms are required for various private small-holder investments in marketing, transport facilities, post-harvesting technologies and quality improvements.[33]

Yet despite implicit subsidies, public liquidity credit is estimated to meet only 15% of the demand for credit by farmers. The other 85% is obtained informally at an interest rate of around 60%. Small farmers, particularly those outside lowland irrigated areas, are especially dependent on high cost, informal sources of funds. And, despite the fact that over 50% of the subsidized liquidity credit goes to sugar production, it accounts for only 3.3% of the value of total crop production in Indonesia. There is also concern that certain subsidized and liquidity credit-financed priority programmes, such as in the major tree-crops sector, may distort the capacity of small-holder producers to become financially viable. These distortions in the credit market, and the general lack of multi-purpose credit at affordable rates with medium- and long-term payback periods, are major constraints on the sustainable development of agricultural lands.

Alternative policies

How can existing policies be improved or replaced? Since producer prices for the major food crops in Indonesia – rice and corn – have generally followed the underlying trend in world market prices, there seems little need to change these. But for upland soya bean and other higher valued upland crops, improved quality and yields may in the long term be a more effective way of increasing farmer incomes than the current practice of maintaining domestic prices well in excess of world levels. There is a particularly strong argument for reducing the very high effective protection rates for vegetables and sugar production, since these are not conducive to improved soil conservation practices in upland areas, and may in fact benefit the richer farmers more than poorer upland farmers.

It may be necessary to continue some restrictive import controls for perennial fruits and animal husbandry products so as to encourage the spread of agroforestry and livestock-based forage systems, particularly in the uplands of Java. But over the long term, Indonesia will need to develop export markets for certain products, such as tropical fruits, which will require a gradual dismantling of protectionist policies. In general, for all export crops vital to sustainable upland development (e.g. coffee, cloves, tea and cocoa), not only does international competitiveness need to be maintained by an effective exchange-rate policy, but monopolistic trading practices must be removed to allow the benefits of improved terms of trade to reach upland small-holders.

Perhaps the major change is most needed in those policies – particularly input subsidies and investment strategies for research, extension and infrastructure – which are still largely biased towards lowland irrigated agriculture, especially rice cultivation. These result in an under-investment of resources in other agricultural areas that are currently absorbing labour and could potentially yield higher growth and incomes. They also artificially overvalue the contribution of the lowlands to agricultural development. Furthermore, high input subsidies encourage wasteful use which is the direct cause of serious environmental problems (e.g. pest outbreaks, over-fertilization) and act as disincentives for proper management of land and water resources. With Indonesia now producing rice surpluses –

resulting in additional high costs for storage of excess stocks and for subsidized exports – there is a case for introducing a phased reduction of these subsidies and reallocating funds towards higher priority agricultural investments for sustainable agricultural development in non-rice growing areas.

Reducing or eliminating input subsidies and reallocating research and extension funds could, in the short term, release US$275 million annually for investment in more sustainable agriculture.[34] Assuming a complete phase out over time of the fertilizer s bsidy and a four-fold increase in both research and extension budgets, this could increase to as much as US$525 million per year. Such funds could be used, very effectively, for:

(1) integrated pest management (IPM), for brown planthopper control, to be gradually extended to IPM for other pests;

(2) increasing the availability of general rural credit, particularly to marginal farmers, at affordable rates and with multiple terms;

(3) research and extension to develop and support new farming systems and land-management techniques appropriate for the marginal (mainly dryland and swampland) sedentary agriculture in the Outer Islands and the uplands of Java, as well as shifting cultivation. This would include the development and dissemination of new varieties appropriate to diverse agro-ecological conditions, research into pest and disease outbreaks, and improvements in small-holder estate crop systems; *and*

(4) investment in: a) further improvements in farming systems for specific agro-ecological zones; and b) improve ments in the physical infrastructure serving these zones, including rural transport, integration of markets, credit facilities, post-harvest technology and processing, and produce quality.

We have discussed the particular case of Indonesia in some detail because it clearly illustrates the manner in which government policies interact with one another to inhibit sustainable agricultural development. As we have tried to demonstrate, policies can be changed, and in ways that not only further sustainability but improve efficiency. The next question is how to translate such policies into workable programmes and projects that which affect the livelihoods of small farms. We discuss this issue in the next chapter.

Notes

1. See, for example, Dunstan S.C. Spencer, "Agricultural research: lessons of the past, strategies for the future", in Robert J. Berg and Jennifer Seymour Whitaker (eds), *Strategies for African Development* (Berkeley: University of California Press, 1986), pp.215–41; and John Mellor, "The changing food situation – a CGIAR perspective", paper presented at the International Centers Week, IFPRI, Washington, DC, 5–9 November 1984; Joachim von Braun and Eileen Kennedy, "Commercialization of subsistence agriculture: income and nutritional effects in developing countries", *Working Papers on Commercialization of Agriculture and Nutrition*, no.1 (Washington, DC: IFPRI, 1986); and FAO Committee on Commodity Problems, 55th Session, *International Trade and World Food Security*, Rome, 21–25 October 1985. For references linking the food- versus export-crop controversy to issues of agricultural sustainability and natural resource management see note 2 below.
2. See, in particular, Edward B. Barbier, "Cash crops, food crops and sustainability: the case of Indonesia", *World Development*, vol.17, no.6, pp.879–93 (1989), pp.1378–87; also Robert Repetto, *Economic Policy Reform for Natural Resource Conservation* (Washington, DC: WRI, 1986); and IIED/WRI, "Food and Agriculture", *World Resources 1987* (New York: Basic Books, 1987).
3. Braun and Kennedy, op. cit., p.1 and table 1.
4. Braun and Kennedy, op. cit. See also U. Lele, "Terms of trade, agricultural growth, and rural poverty in Africa", in J.W. Mellor and G.M. Desai (eds), *Agricultural Change and Rural Poverty* (Baltimore: Johns Hopkins University Press, 1985).
5. World Bank, *World Development Report 1986* (Washington, DC: World Bank, 1986), p.79.
6. J. Boesen, E. Friis-Hanses, S. Garset, M. Speirs, R. Odgaard and H.M. Raynborg, *Research on the Increasing Vulnerability of Peasant Farming Systems, Their Resource Basis, and Food Production in East Africa* (Copenhagen: Centre for Development Research, August 1987). See also E.B. Barbier, "Natural resources policy

and economic framework", Annex 1 in James Tarrant *et al.*, *Natural Resources and Environmental Management in Indonesia* (Jakarta: USAID, 1987); Robert Chambers, *Sustainable Livelihoods, Environment and Development: putting poor people first*, Discussion Paper 240 (Brighton: Institute of Development Studies, University of Sussex, 1987); R. Dennis Child, Harold F. Heady, Wayne C. Hickey, Roald A. Peterson and Rex D. Pieper, *Arid and Semiarid Lands: Sustainable use and management in developing countries* (Morrilton, Arkansas: Winrock International, 1984); and Gordon R. Conway, Ibrahim Manwan and David S. McCauley, "The Development of Marginal Lands in the Tropics", *Nature*, vol.304, p.912.

7. Chambers, op. cit., pp.2–3.
8. Robert Repetto, *World Enough and Time: Successful strategies for resource management* (New Haven: Yale University Press, 1986).
9. R. Chambers, op. cit., p.4.
10. ibid.
11. Roland Bunch, "Case study of the Guinope Integrated Development Program", paper presented at Conference on Sustainable Development, IIED, London, 28–30 April.
12. See, for example, Paul Harrison, *The Greening of Africa: Breaking through the battle for land and food* (London: Paladin, 1987).
13. See Richard Ackerman, Edward B. Barbier, Gordon R. Conway and David W. Pearce, "Environment and sustainable economic development in Indonesia – an overview report" (draft) (London: IIED, 1987); and World Bank, *Indonesia – Java Watersheds: Java Uplands and Watershed Management*, (draft) (Washington, DC: World Bank, November 1987).
14. Jeremy J. Warford, *Environment, Growth, and Development*, Development Committee Paper no.14 (Washington, DC: World Bank, 1987), p.19.
15. World Bank, *World Development Report 1986* (Washington, DC: World Bank, 1986), pp.67–8; and Kevin M. Cleaver, *The Impact of Price and Exchange Rate Policies on Agriculture in Sub-Saharan Africa*, World Bank Staff Working Papers no.728 (Washington, DC: World Bank, 1985).
16. The above discussion has been based on Per Pinstrup-Andersen, "Food prices and the poor in developing countries", *European Review of Agricultural Economics*, vol.12: pp.69–81; and John W. Mellor and Gunvant Desai (eds), *Agricultural Change and Rural Poverty: Variations on a theme by Dharm Narain* (Baltimore: Johns Hopkins University Press, 1985).

17. World Bank, *Indonesia – Agricultural Policy: Issues and options* (Washington, DC: World Bank, 1987).

18. Edward B. Barbier, "Natural resources policy and economic framework", Annex 1 in James Tarrant *et al.*, *Natural Resources and Environmental Management in Indonesia* (Jakarta: USAID, 1987), pp.1–31.

19. The following is based on Ackermann *et al.*, op.cit.; and World Bank, *Indonesia – Java Watersheds*, op. cit., ch.7.

20. Bruce Glassburner, "Macroeconomics and the agricultural sector", *Bulletin of Indonesian Economic Studies*, vol.21, no.2 (April 1985).

21. Although only 10% of cassava is exported, 97% of exports are to the EEC. See Faisal Kasryno, *Analysis of Trends and Prospects for Cassava in Indonesia* (Bogor, Indonesia: Center for Agroeconomic Research, Agency for Agricultural Research and Development, 1987).

22. Studies of upland farming on Java have revealed that the crucial economic factors influencing upland farmers' decisions to adopt soil conservation measures include:
    (a) profitability of the farming system
    (b) nature of the farming system
    (c) economic status of farmer
    (d) relative importance of on-farm to off-farm income
    (e) security of land tenure
    (f) mode of tillage
    (g) distance of land from roads and farmers' homes.
    See, for example, Brian Carson, with the assistance of the East Java KEPAS Working Group, *A Comparison of Soil Conservation Strategies in Four Agroecological Zones in the Upland of East Java* (Malang: KEPAS, 1987), pp.32–7. See also Edward B. Barbier, *The Economics of Farm-Level Adoption of Soil Conservation Museums in the Uplands of Java* (London: IIED, 1988).

23. World Bank, *Indonesia – Java Watersheds*, op.cit., ch.4.

24. World Bank, *Indonesia – Agricultural Policy: Issues and options*, op. cit., p.48.

25. World Bank, *Indonesia – Agricultural Policy: Issues and options*, op. cit., Technical Annex E, table 1.

26. Bambang Irawan, "Executive summary: marketing analysis for dryland farming development in Citanduy river basin" (Ciamis, Indonesia: USESE, 1986).

27. GOI/Ministry of Agriculture/DG of Estates, "Coconut processing and marketing", *Feasibility Study, Smallholder Estate Crops Development Project*, vol.8 (1987), pp.8–18.

28. World Bank, *Indonesia – Agricultural Policy: Issues and options*, op.cit., ch.7.
29. World Bank, *Indonesia – Java Watersheds*, op.cit., pp.17–18.
30. E.B. Barbier, "Natural resources and economic policy framework", op.cit., pp.1–16.
31. World Bank, *Indonesia – Java Watersheds*, op.cit., ch.7.
32. Peter E. Kenmore, "Status report on integrated pest control in rice in Indonesia with special reference to conservation of natural enemies and the rice brown planthopper (*Nilaparvata lugens*)", FAO Indonesia, 13 October 1986.
33. World Bank, *Indonesia – Java Watersheds*, op.cit., ch.7.
34. This amounts to US$150 million from a reduction in the fertilizer subsidy, US$25 million and US$40 million from the abolishment of the pesticide and sugar credit subsidies respectively, and US$30 million each for reallocation of research and extension funding.

5. Farms and Livelihoods

In the preceding two chapters we have illustrated the importance of international and national policies in the promotion of sustainable development. Indeed, it could be argued that without such policies in place there is little chance of sustainable agriculture being achieved. No amount of research and appropriate extension will persuade farmers to conserve resources if powerful economic incentives are driving them in the other direction. Nevertheless, sustainable agriculture ultimately depends on the individual, day-to-day actions of millions of farmers and their families, pursuing a variety of strategies aimed at securing their livelihoods.

While programmes of research and analysis have a crucial role to play in policy reform, it is inevitable that the major proportion of development funding and effort over the next decade will go to executing projects whose primary focus is an identified group of farmers in a particular region, watershed, district or even village. The challenge is how to help them satisfy their needs in ways that are efficient and sustainable. In this chapter we examine the dimensions of the problem.

Farming systems

With the shift in focus in the 1970s – from the homogeneous, well-endowed and controlled environments typical of the green revolution lands, to the needs of farmers in more marginal and heterogeneous environments – came a significant change in research and extension emphasis. Because of the greater diversity and complexity of farms in resource-poor environments, it became apparent that these farms had to be understood as whole systems and not simply as collections of individual agricultural commodities. Moreover, the research was not to stop there, the new systems-understanding was to be used to develop new technologies that were appropriate in a system context. The result was the development of the approach known as Farming Systems Research and Extension(FSR&E)[1].

There is much discussion in the literature as to the precise nature and remit of FSR&E, but most practitioners would agree that FSR&E can be distinguished from traditional agricultural research and development by the following characteristics:

(1) a systems framework of analysis rather than a commodity-based approach;
(2) the explicit incorporation of social science perspectives and methodologies, and attention to both biophysical and socio-economic constraints to production;
(3) an attempt to obtain participation by farmers in the research and development processes, especially through on-farm trials; *and*
(4) the utilization of the small farm as the unit of analysis.

In nearly 20 years, it has evolved in a number of different directions, in the hands of both the international agricultural research centres (IARCs) of the Consultative Group on International Agricultural Research (CGIAR), and of national research institutes and universities. Matching this variety of approaches has been a mixed record of success and failure, in terms of providing significant benefits to farmers on resource-poor lands.

Technology push and farmers' needs pull

In general the outcomes have depended on who has conducted FSR&E and for what purpose, both explicit and implicit. There have been many classifications of FSR&E but one which emphasizes goals and approach is the distinction between "Technology push" and "Farmers' needs pull" FSR&E.[2]

The former arises from the desire of technology innovators to see how well their innovations are adopted by farmers in the field. It is the most immediate and obvious response to the "gap" between performance in farmers' fields and on international or national research stations. The focus is "on-farm research" (OFR) in which new technologies such as alley-cropping, or disease-resistant varieties, are tried out in farmers' fields in the hope of better appreciating the environmental and socio-economic constraints to adoption. Trials are placed on plots in farmers' fields and under regimes that range from researcher-managed to farmer-managed. The potential for learning from this experience is considerable, but all too frequently the desire of the technology innovators to see success progressively reduces genuine farmer-participation (in some cases farmers are simply left with the mundane task of weeding the plots; in others, when farmers "fail," the researchers take over). The environmental constraints are often illuminated, but the socio-economic constraints continue to be ignored.

The "Farmers' needs pull" form of FSR&E is, at least on paper, radically different. Here the starting point is not new technology but the analysis of existing farming systems, *in situ*, to determine needs, problems and constraints to which subsequent technological innovation is directed. The Centro Internacional de Mejoramiento de Maiz y Trigo (CIMMYT) project in East and Southern Africa, although having a strong "maize-systems" orientation, has tried to follow this approach.[3] Its procedure, which is termed "on-farm research with a farming systems perspective" (OFR/FSP) consists of the following steps:

(1) diagnostic survey;
(2) the identification of farmers' needs;
(3) the search for appropriate technology;
(4) testing via on-farm trials; *and*
(5) recommendation for adoption.

The steps are followed sequentially but the approach is intended to be iterative, a recurrent analysis of the farmer's system permitting continuous learning and adaptation (this emphasis often results in the alternative name of "adaptive on-farm research").

Another major "farmers' needs pull" programme has been developed under the rubric of agroecosystem analysis (AEA) rather than FSR&E, and initially focused on Southeast Asia (see Appendix).[4]

It is not easy to summarize the lessons from this range of FSR&E experience. In the donor community as a whole there is some degree of disillusionment with FSR&E, partly due to quite unrealistic expectations of rapid pay-offs. As a recent survey by the CIMMYT project reveals, it is difficult to point to clear cases of farmer adoption of technologies developed under the FSR&E rubric. But this, of course, is not surprising given the relatively short time that FSR&E programmes have been in operation, and the quite radical reorientation of people and institutions that successful FSR&E entails. It is easier to identify cases where technologies have been modified and research priorities changed in response to a better understanding of constraints.

Perhaps the most conspicuous result has been quite dramatic changes in the perceptions of farming systems and farmers' needs among agricultural researchers in a number of universities and national research institutes. This has been most pronounced where demand for FSR&E has grown out of a frustration with conventional research approaches and a genuine desire to understand the complexities of small farmers and their needs. In general, the response to this demand has been most successful in the absence of direct involvement by the IARCs. Where IARCs have been productively involved, as in the notable case of CIMMYT in Africa, they have not stuck to their mandate, and have been open to a farmers' needs approach. In the CIMMYT programme the leadership has significantly been from socio-economics rather than agronomy and the maize–wheat mandate has not been dominant. The International Center for Agricultural Research in Dry Areas (ICARDA) has had a moderately successful impact for similar reasons – it has also had the advantage of including a livestock element in its mandate which helps to ensure that FSR&E does not remain narrowly focused on cropping systems. In general the broader the mandate of the IARC the more capable it is of responding to a "farmers' needs" approach.

Narrow, single crop mandates tend to generate a "technology push" form of FSR&E.

Livelihoods

Even where FSR&E is successful in terms of meeting farmers' needs, it still remains only a partial approach to the problem of attaining sustainable agriculture. This is primarily for two reasons. First, with surprisingly few exceptions, developing country farmers, particularly in resource-poor environments, do not rely exclusively on farming. Their aim is to secure a livelihood for themselves and their families and to achieve this they usually pursue a range of productive activities, only some of which involve crop or animal husbandry. Second, with even fewer exceptions, farms do not exist in social isolation but are integral components of communities, whose institutions, customs and systems of rights and obligations determine much of what farmers can and cannot do.

A livelihood is usually defined as the means of securing a living, but this brief definition obscures a concept which is complex in both theory and practice.[5] Encompassed in a livelihood is the totality of resources, activities and products which go to securing a living. It relies on ownership of, or access to, resources, and on access to products or income-generating activities. A livelihood is measurable in terms of both the stocks – that is the reserves and assets – and the flows of food and cash.

The ways in which a rural livelihood may be obtained are almost innumerable. It may depend on land on which crops or livestock are husbanded; or on natural resources – timber, fuelwood, wild plants, fish and other wild animals – which may be harvested; or on opportunities for off-farm employment; or on skills employed on the farm in manufacture of handicrafts; or, most commonly, on some combination of these. In practice, rural families decide on livelihood goals and then determine the optimal mix of activities, depending on their environmental and social circumstances and the skills and resources at their disposal.

Few comparative studies of livelihoods have been carried

out, but a particularly illuminating livelihood analysis has been conducted for four Amerindian groups in Central Brazil who practise a combination of gardening, hunting and fishing.[6] Table 5.1 shows the yields per person hour for these different productive enterprises, together with the time spent on each, which is a measure of how the people value these productivities.

In the case of fishing the relationship between yield and effort is fairly straightforward; the higher the productivity the greater the time spent fishing. For gardening and hunting though the relationship is more complex. The gardens of the Mekranoti, for example, produce far more calories than they could possibly eat; the excess is stored as insurance against bad years, or kept to feed visitors from outside, and this appears to be why they spend less time gardening than the others. Hunting is also more productive for the Mekranoti, but here they devote more time to hunting than the other groups. In consequence they eat a large amount of animal protein, possibly because they have the time to do so and appreciate the quality of a high-protein diet.

The Kanela have a high population density and live in a much poorer habitat. Hunting and fishing are very unproductive and they spend a relatively large amount of time gardening, concentrating on protein-poor manioc to provide the calories they require. They also spend more time than the other groups in producing handicrafts for sale or working for wages, in order to make up the protein shortfall and to satisfy other requirements. The Xavante lie between these two extremes.

The Bororo also live in a poor environment for gardening, but fish production is high. They sell some fish for high-calorie foods, but the market is far away and they get a low price. Fishing is also very hard work and they suffer from a high rate of illness and invalidism. They also have the lowest ratio of dependent children which possibly gives them a lower incentive to increase production.

Such an analysis may seem of purely academic interest, but it clearly reveals some of the practical constraints to development. The sustainability of these livelihoods depends on their diversity, and potential innovations – such as new crops to improve the vegetable protein intake of the Kanela, or the calorie intake of the Bororo – must not conflict, in terms of labour demand, with the existing patterns of profitable activity. It can be argued that livelihood analysis is an essential prerequisite for sustainable development interventions, yet to

Table 5.1: Productivities of four Brazilian Amerindian livelihoods

	Mekranoti	*Xavante*	*Bororo*	*Kanela*
Gardening				
Ave. yield (10^3 kcal) per person hour	17.6	7.1	1.5	5.1
Hrs gardening per day per adult	1.21	2.09	1.44	2.50
Hunting				
Yield (kg dressed meat) per person hour	0.69	0.40	0.20	0.11
Hrs hunting per day per adult	0.87	0.47	0.09	0.55
Fishing				
Yield (kg dressed fish) per person hour	0.20	0.40	0.68	0.05
Hrs fishing per day per adult	0.21	0.44	0.50	0.13
Production and consumption				
Produce (10^3 kcal) per person per day	21.3	14.8	2.20	2.80
Vegetable protein (g) per person per day	89	138	24	53
Animal protein (g) captured per person per day	63	37	44	7
Animal protein (g) consumed per person per day	72	28	81	18

Source: D. Werner, N.M. Flowers, M.L. Ritter and D.R. Grass, "Subsistence productivity and hunting effort in native South America", *Human Ecology*, vol. 7, 1979, pp. 303-15.

date few such livelihood analyses have been carried out for rural households.

Off-farm income

Another important outcome of livelihood analysis is an appreciation of the role of off-farm income in resource-poor lands. This is well illustrated by the case of two villages in upland Java, both situated on limestone soils which have suffered from severe erosion.[7] Table 5.2(a) shows the agricultural production in each village.

Merden is the poorest village. It has smaller landholdings, steeper slopes and greater erosion. The terraces are poor, only one crop a year is grown, and there are goats but no cattle. By contrast the people of Bunder have larger landholdings and access to a nearby state forest. Their terraces are good, they grow two crops a year, and own cattle which are fed on elephant grass grown along the terraces and fodder trees introduced by a

Table 5.2(a): Productivity of two upland villages in Central Java

| | *Crop production (kg/ha/year)* | | | | *Production value* | |
	Maize	*Cassava*	*Rice*	*Peanuts*	*kcal/ha/yr* (000)	*Rp/ha/yr* (000)
Bunder						
<0.5ha	485	1,770	440	1,020	5,240	332
>0.5ha	270	1,025	415	655	3,565	233
Mean	348	1,295	425	790	4,177	269
Merden						
<0.5ha	795	3,790	–	–	7,000	97
>0.5ha	450	1,740	–	–	2,515	49
Mean	700	3,200	–	–	6,010	83

Source: P. van de Poel and H. Van Dijk, "Household economy and tree growing in upland central Java", *Agroforestry Systems*, vol. 5, 1987, pp. 181-4.

government regreening programme. Manure is applied to the crops.

Yet in neither village is food production sufficient for subsistence. Average family size in both villages is about five adults with a minimum food requirement of about 3.6 million kcals per year. In Bunder the average production per household is 2.5 million kcals per year and in Merden it is 2.2 million kcals per year. Both villages survive because of off-farm income: government jobs, carpentry, trade and the selling of charcoal and wood in Bunder; and wage labour, mining, carpentry, trade and palm-sugar production in Merden. The sources of cash income are given in Table 5.2(b).

As in the previous example, agricultural innovation has to take into account the returns to agriculture relative to those from other forms of productive activity, in this case off-farm labour. Moreover this comparison has to be made in terms of both the long-term and short-term trade-offs – the relative productivities have to be assessed for sustainability as well.

In the case of these Javanese villages most of the outside income is earned locally, but frequently very large sums of money are

Table 5.2(b): Estimated cash income and relative importance (%) of three sources of revenue in generating cash income (livestock sales not included) for two upland villages in Java

	Merden		*Bunder*	
	<0.5ha	*>0.5ha*	*<0.5ha*	*>0.5ha*
Total income per house-hold (Rp 1,000)	221	250	769	481
Income per capita (Rp 1,000)	49	41	175	82
Off-farm activities	95%	90%	93%	70%
Crops	2%	4%	4%	16%
Wood/fruits	3%	6%	3%	14%

Source: P. Van de Poel and H. Van Dijk, "Household economy and tree-growing in upland Central Java", *Agroforestry Systems*, vol. 5 (1987), pp.169–84.

are sent back to rural communities by individuals who have temporarily or permanently migrated to the major urban centres or even overseas. In recent years, rural livelihoods in Mexico have significantly benefited through such remittances from the United States and, similarly, in southern African nations though remittances from South Africa, and in many Asian countries from Saudi Arabia and the Gulf States. Even quite remote villages in the mountains of Pakistan or small islands in the Philippines are tied in this way to the economies of the oil-rich Arab states. The total amounts involved in world-wide remittances are not known; they probably exceed official development assistance. More important, very little is yet known about how such remittance moneys are used. To what extent are they spent on food, on consumer goods, on education, or on investment in land and long-term agricultural production?

Security

There does seem to be growing empirical evidence, however, that farmers are likely to invest such moneys and, indeed, any other savings, in activities that have a significant long-term pay-off. Much depends on the circumstances. For the very poor , sheer survival is the priority, and however much they may wish to, people find it difficult to take the long-term view. For the poor, though, once basic survival is assured, and given safe and secure conditions, there appears to be a strong propensity to stint and save when the opportunity presents.[8] Security, in one form or another, seems to be the key to encouraging investment of labour and funds in resource conservation and enhancement.

An example of such investment occurs in Kakamega and other districts in the Western Provinces of Kenya. There, rapid population growth has produced densities of the order of 700 persons per km^2, yet contrary to the conventional wisdom that such environments should be suffering from acute deforestation, the hillsides are covered with trees, although they are planted rather than natural. On some farms, as illustrated in Figure 5.1, the trees are essentially a kind of subsistence crop, being sold to obtain basic food during the lean months of the year; on other

Figure 5.1: A modern half-acre farm, Kakamega, Western Kenya

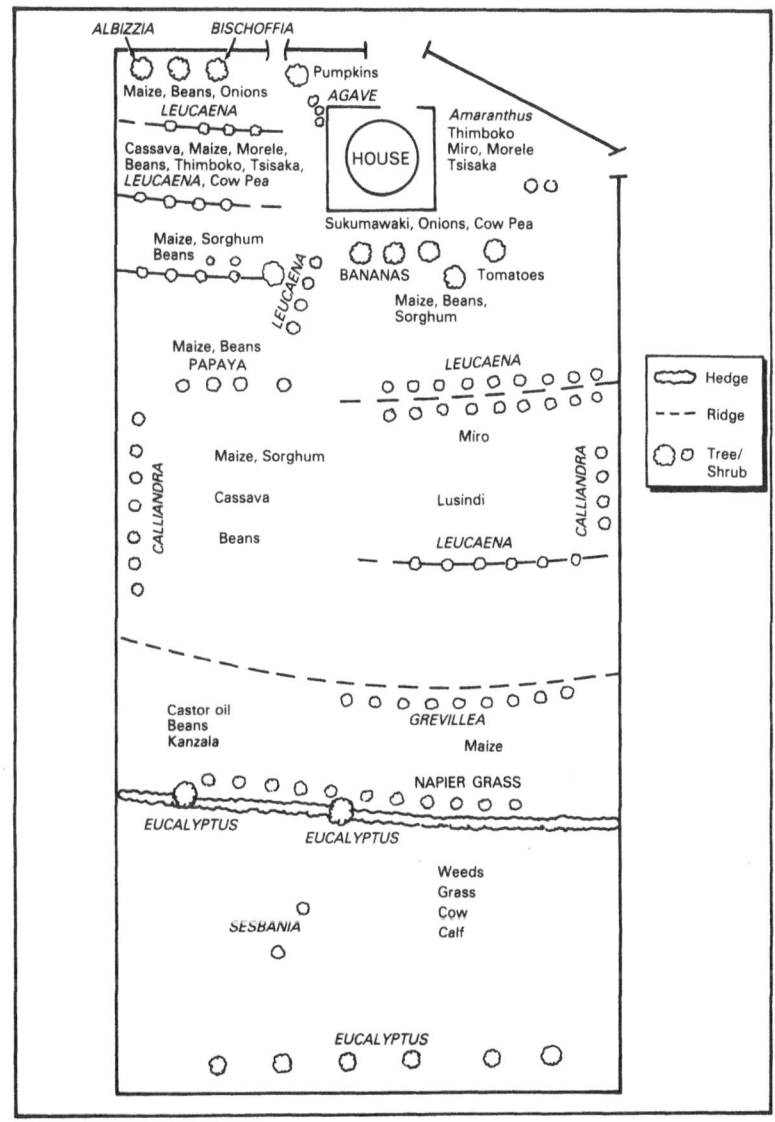

farms they are savings for the inevitable bad year or the expense of a wedding or funeral; on yet others they are grown for cash and, in particular, for the regular payment of school fees. In the latter situation the sustainable investment is in the long-term future of the children. The clue to this remarkable pattern of investment appears to be that the farmers have secure ownership of their land, however small it may be. The example is not an isolated one: similar patterns of tree planting under conditions of security have been reported from India and Haiti.[9]

Governance

The equation of land security and investment for sustainability is probably too simplistic, however. Most rural communities are characterized by quite complex systems of rights and obligations to land and other property, and to resources in general. The outcomes in terms of long-term resource conservation may not be readily predicted if only one aspect is considered. Thus ownership of land may not be sufficient to encourage tree planting or terracing if other rights are not also secured or other conditions are not satisfied.

For example, evidence from Java suggests that economic status and security of tenure of upland farmers are important determinants for adoption of conservation packages on shallow, poor, erosion-prone soils. On the other hand, on deeper, more fertile soils which are equally subject to erosion and yet have little short-term loss of productivity as a result, even better-off farmers with security are less likely to practise soil conservation as they cannot perceive a tangible gain. They apparently do not appreciate the long-term degrading effect of erosion and indeed they are not the principal sufferers: these are the farmers downstream whose irrigation systems become silted. Upstream farmers have no traditional obligations to those downstream.

Evidence from northern Thailand also indicates that these links between land insecurity and environmental degradation are complex. Results of a study for Nakhon Ratchasima, Lop Buri and Khon Kaen suggest that land security is important for land investment, but this is mainly due to the ability of secure owners to obtain institutional credit. Another survey of villages in northern Thailand seems to contradict the claims that land ownership is a necessary pre-condition for sound resource management. In the villages surveyed, lack of labour and money appear to be the two greatest impediments to soil conservation, whereas insecure land tenure appears to be hardly significant. This is explained by the continued availability of new lands, the security of some farmers' tenure, and the quasi-legal recognition of rights of shifting cultivators to clear and till land.[10]

The common property problem

Similarly it is often believed that the common ownership of resources, such as land, is the major cause of resource degradation in agriculture. That is, each user of the commonly-owned resource may maximize his or her "share" at the expense of the externality impacts of any resulting degradation on others, or on his or her future use.[11] Rangeland management in tropical semi-arid areas is often assumed to suffer from the "common-property problem". Under communal land ownership each individual can maximize his or her "share" of pasture by increasing his or her own livestock. This is presumed to lead in aggregate to collective overstocking, producing degradation of the pasture, something which, conversely, the individual cannot avoid by unilaterally limiting his or her herd. Many traditional pastoral livestock systems, however, have evolved highly organized controls on the use of common-property land, with sanctions by the community against individual over-exploitation. Where such traditional controls exist, resource sustainability is not a problem.

Nevertheless, many rapid changes in pastoral areas may be transforming common-property resources into "open-access" resources, where traditional communal management institutions no longer apply. These include:

- the introduction of technological changes, such as trucks to transport people and animals; and government interventions, such as declaring grazing land to be public property
- increasing competition between different ethnic groups,

some perhaps dislodged from their traditional grazing areas

- growing stratification of income and wealth, alongside erosion of social cohesiveness and reciprocal caring arrangements which, in particular, cause conflicts among large and small livestock owners
- rapid population growth and the abrogation of traditional lands which, in particular, confines traditional migratory patterns during dry seasons and thus increases grazing pressure.[12]

Rights and the resource base

The general conclusion from these examples is that without careful empirical study of actual relationships at the grassroots level, the tendency may be to misinterpret the impacts which institutional arrangements for land ownership, rights of access and tenancy have on environmental degradation. The analysis is further complicated by the fact that few Third World rural households are bound by one set of institutional property rights arrangements; many own a little land, rent in a little more, do some farm labour for other, bigger owners and even have some rights over certain commonly-owned resources. Moreover, security of access to land affects poverty and resources differently in favoured and marginal agricultural regions. Under irrigated, improved or modern farming conditions leading to high net returns per hectare, access to even a little bit of land, despite being associated with larger household size, reduces the probability of poverty in an average year. On very resource-poor lands, however, farmers with small and even middle-size holdings tend to be only marginally better off than landless labourers.

For example, data from western India's semi-arid lands indicate that the incidence of poverty hardly changes as owned landholding rises from about 0.5 acres to 7.5 acres. One explanation may be that such marginal lands, under the present state of knowledge, do not yield enough to pay for capital investments and are thus worked with much labour and draught-animal power relative to capital. Thus the small farmer is

little better-off than the landless labourer, because the latter
is less readily displaced by capital; and until recently with the
increasing loss of common grazing rights, both groups may
be less worse-off than the middle farmer, because livestock
is more equally distributed than land. In general, institutional
property-right arrangements may be less determined by notions
of "optimum size" of landholding than by relations among
groups of producers and by farmers seeking to adjust land
to other resource endowments. This explains, for example,
why imperfections in the market for oxen hire are frequently
associated with shifts of land from owners with few oxen to
operators with many.[13]

Households

The final important point about livelihoods is that they rarely
relate to an individual. Most rural livelihoods are those of
family households, comprising men and women, and further
distinguishable as adults, children and the elderly. In some cases
it is appropriate to consider the collective livelihood of extended
families, in which case the household includes aunts, uncles,
cousins and people of even more distant kin relationship.

Each household not only determines the optimal mix of on-
farm and off-farm activities which goes to make up its livelihood
but also allocates these tasks among its members. In recent
years the interest among development workers in gender issues
has provided a growing knowledge of how these decisions are
made, their consequences and their implications for sustainable
development.[14]

The complexity of both inter- and intra-household decision-
making is well illustrated by studies of the Tswana people in
Botswana.[15] Like many other rural communities in southern
Africa their livelihoods are a combination of crop production,
stock raising and off-farm employment, including wage labour.
Yields of sorghum and maize are extremely low (250 kilogrammes per hectare (kg/ha) and few families have the 8 ha or
more that would be necessary for a subsistence diet. Because of
good marketing and high prices, cattle raising for beef is a better

income earner, but cattle holding is very skewed. In particular, a significant proportion of the population does not have cattle for ploughing, or has unreliable access to cattle, and hence the timeliness of ploughing on which good harvests depend is jeopardized. Remittances, particularly from South Africa, are crucial and, indeed, the poorer 50% of households rely on such remittances as their primary source of income.

Such a livelihood system creates many problems. First, if men are away from home, who looks after the cattle? This is solved by the men taking turns at being away, a process which may span several generations. For example:

> In generation A, a young man goes as a migrant labourer to South Africa. He remits moneys to his parents and sister at home, where they grow food crops and maintain a small herd of cattle. After the parents die, the sister, who remains unmarried but has children, maintains the fields, and her growing sons herd the cattle. In generation B, the now elderly, former migrant labourer keeps the cattle belonging to his dead sister's sons, who now work as migrant labourers. The labour of the late sister on the fields and her sons in herding has helped build the herd which, on the death of now-old man, will be divided among his children and his sister's children."[1]

Another problem is that of getting access to draft animals. This is solved by exchanges of human labour, for the most part female labour, and this is one of the areas in which women play a crucial role as negotiators and managers. The studies have also shown that the stereotypes of sexual divisions of labour are too simplistic. For example, women do plough, and they play a more direct role in cattle-keeping than has been supposed. Surveys have also shown that in a great many areas husbands and wives make joint decisions, and indeed, because of their managerial role, women can be regarded as the "lynch pin" in the diversified activities of the Tswana household. These and other similar studies clearly show that development interventions in terms of, say, new crop varieties and breeds of cattle, are only likely to be acceptable and sustainable if they take into account the effects they may have on these complex inter- and intra-household relationships.

Participation

The complexity of farms, households and livelihoods is a daunting prospect for the designers and managers of development projects. It would seem impossible to obtain, in a reasonable period of time, sufficient understanding of the dynamics and trade-offs involved in any particular situation so as to identify the real opportunities for sustainable development. However, there are solutions to the problem. Part of the answer lies in the use of relatively quick, multidisciplinary techniques of analysis and appraisal, some of which are described in the Appendix. But perhaps the most compelling solution lies in involving farmers and their families themselves in the design, selection and management of innovations intended to improve their lives.[16] The logic is fairly clear; they, after all, live with the complexity of their physical and social environment and with the need to make difficult trade-off decisions on a day-to-day basis. Moreover there is good evidence, from many parts of the world, that the understanding by farmers of their own situations is deeper and more insightful than has been commonly supposed.

First, there is abundant evidence of the capacity of farmers to respond positively to innovations and new opportunities if they can see the benefits in both the short term and the long term. A case in point is the response of farmers to new rice varieties in India. Much effort by breeders at the national breeding centres in India and at the International Rice Research Institute (IRRI) has gone into breeding new high-yielding varieties, but when released many have not been adopted by farmers, for one reason or another.[17] Yet there have been strains, rejected during trials by the breeders, which have won ready acceptance from the farmers. The variety Mahsuri, for instance, was rejected by breeders in the All-India Coordinated Tests because it tended to lodge, but farmers who saw it growing liked its positive qualities – its semi-tall habit, high tillering, heavy panicle, ease of milling and grain quality – and obtained seed for their own farms. Today it is the third most popular variety in India. Another variety from IRRI, designated IR24, was rejected by breeders because it flowered poorly under low temperatures when it was late planted. But one farmer obtained a sample, grew it on his own farm and saved the seed from those few plants which did flower well. He called this new variety "Indrasan" which became widely grown in the states of Uttar Pradesh and Haryana and proved to be resistant to a major attack of white-backed planthopper which occurred in 1985.

Second, farmers are experimenters and innovators in their own right, and literature has many examples of quite sophisticated experimentation among farmers. Paul Richards describes the varietal trials that the Mende people of central Sierra Leone carry out as a matter of course.[18] They acquire new rice material by begging or buying from friends and visitors and sometimes in the course of their travels to other parts of the country. Such material is first tried out near the farm hut; it is harvested with a knife, panicle by panicle so that the best grains can be retained. Full-scale trials are then undertaken in the "seep zone". This is the marginal land between the uplands and the swamps below. Depending on the behaviour of the new material the farmer does further trials up or down the escarpment. In each case he makes accurate input-output measurements, using for example the same sized calabash to record the amount sown and harvested. Paul Richards also reports farmers in Nigeria, worried by reduced fallows and declining fertility, experimenting with new intercropping systems, for example water yams and white yams. For these farmers experimentation is seen as an end in itself, which might or might not have practical consequences.

What is often forgotten by development experts is that much of the agricultural innovation in the West, particularly during the so-called agricultural revolution of the eighteenth century, came from the farmers themselves. The Norfolk four course rotation, Jethro Tull's seed drill and the plaster and clover system of Pennsylvania were all the products of experimenting farmers. There is no reason to suppose that such a capacity is not present among the peasantry of the developing countries today.

Until recently farmer participation in development projects has been largely confined to on-farm trials. Far too often these have simply been reflections of a technology push FSR&E approach (as previously described), in which farmers are little more than experimental field assistants labouring on experiments on their own land. Nevertheless, there have been a number of conscious attempts to increase the true level of participation.

Farmer participation in research

In India, when it was realized that breeders' and farmers' objectives were often at variance, farmers were enlisted in conducting and assessing the trials. Advanced on-station rice lines were matched as closely as possible with the traditional varieties the farmers were growing[17]. Participating farmers were then encouraged to grow the traditional and the experimental lines alongside each other in split plots on their farms. As the rice matured the farmers visited all the plots, in a group, and gave their opinions on the relative performances of the varieties. They also later assessed the harvesting, threshing, milling and cooking qualities of the grains. In virtually all cases the improved lines outyielded the traditional varieties and were pronounced superior in several respects. The enthusiasm of the farmers for this approach was such that they asked for it to be extended to lowland rice and to the winter crops of wheat and barley. From the researchers' point of view, apart from the satisfaction of producing material that was clearly acceptable, this process short-circuited a normally lengthy and costly sequence of screening procedures which has become the norm for breeding programmes.

Further examples are the experiments in farmer participation conducted by the Centro Internacional de Agricultura Tropical (CIAT).[19] These, like the Indian project, were begun because scientists began to realize that the varieties developed according to research station criteria are often not automatically accepted by farmers, and sometimes farmers adopt what researchers consider as inferior varieties.

In the programme, CIAT anthropologists and agronomists attempted to involve farmers as integral components in the whole chain of events from diagnosis, problem-definition and design of experiments to exploration of potential improvements. They began by asking farmers to rank a wide range of bush bean grains, and compared these with the ranking of the breeders. There were significant differences both between the breeders and the farmers, and between men and women, the latter choosing, for example, smaller grains which were better fla-voured, while the men – with an eye on marketing – showed a preference for the larger grain types. Expert farmers were then selected by the community to conduct the trials on their farms and were involved in the choice of what to evaluate. They also carried out the evaluations and expressed their final rankings of the varieties on the basis of the trial. Table 5.3 indicates the relative rankings after the trials of three bean varieties and the reasons the farmers gave for the rankings.

CIAT has also successfully established "innovators' work-shops" – groups of farmers who are able to design and evaluate experiments. In one case, involving snap beans, they suggested a trial to address the problem of lack of stakes for climbing snap beans. Their solution was a rotation with tomatoes which utilized the residual tomato fertilization and standing tomato stakes to support the beans. The trial was carried out and the same group of farmers reconvened in the field for its evaluation. They established a set of criteria, including yield, disease condition and various quality characteristics of the bean, and on these criteria agreed on two varieties which were out-standing.

Sukhomajri

Livelihood and household analysis, participatory research, secur-ing rights, and governance, are all ingredients of sustainable livelihood development. As yet, though, there are few examples of successful projects where all or most of these are present. In conclusion to this chapter we present two such projects, drawn from a conference held in London in 1987 that was expressly aimed at examining successful case studies of sustainable development.[20]

The first of these is the relatively well-known Sukhomajri

Table 5.3 Examples of farmers' reasons for selecting or rejecting bush bean varieties, obtained from farmer evaluations of on-farm trials

	Example 1	Example 2	Example 3
Variety	BAT–1297	A–486	ANTIOQUIA BL–40
Breeders' ranking	10th place (least acceptable)	2nd place	5th place
Farmers' final ranking	2nd place	6th place	8th place (least acceptable)
Reasons for choosing	– "although its grain is small and price is lower, it is still profitable" – "because it is high yielding" – "because flavour is good" – "because it is resistant to disease and pests" – "because it withstands drought" – "it germinates better than other varieties" – "it is good for consumption purposes because it swells to a good size when cooked, it yields in the cooking pot"	– "the size and colour of the grain is very nice when freshly harvested" – "it yields well and is delicious to eat" – "it is early"	– "because it yields well"
Negative aspects	– "the grain is very small" – "it is a later variety than the local one, Calima"	– "it is very quickly infested by storage pests" – "a short time after harvest the dried grain changes colour and is difficult to market"	– "the grain is very variable in colour which makes marketing difficult" – "it is a very bushy or sprawling plant and in the rainy season is much affected by disease, also type of plant makes weeding difficult" – "it has a lot of small pods or immature pods at harvest time" – "it yields well, but a lot of the beans are no good, some are rotten, others are still green" – "it is very late" (compared to the local variety" – "it requires more care, (fumigation) because the plant is large and bushy"

Source: Jacqueline A. Ashby, Carlos A. Quiros and Yolanda M. Rivera, "Farmer participation in on-farm trials", *Agricultural Administration (Research and Extension) Network*, Discussion Paper 22, (London: Overseas Development Institute, 1987).

project in India.[21] Sukhomajri lies in the Shivalik hills, not far from the city of Chandigarh in north-western India. The hills, once heavily wooded, have suffered from increasing human and livestock populations, leading to overgrazing and severe erosion. Near Sukhomajri barely 5% of the uplands had vegetation cover in the 1970s, and erosion rates of 150–200 tonnes/ha were not uncommon. In 1958 a dam had been constructed, creating a lake to serve the city of Chandigarh, but by 1974 over 60% of the lake was filled with sediment.

To protect the lake, the authorities first tried to persuade the gujjar herdsmen of the village to stop using the hills for their cattle and goats, but with little result. The breakthrough came when it was decided to build a small earthen dam in the hills to provide water for the village itself and then to stabilize the catchment of this dam with contour trenches, check dams and the planting of vegetation. The stored water was used to irrigate nearby fields and farmers were provided with subsidized seed and fertilizer. Yields were greatly increased, but the farmers who did not benefit continued to use the hills for grazing. It was then that the villagers collectively proposed that more small dams should be created so as to extend the irrigation system; they also suggested the creation of a water-users' society, based on the principle of equity, to manage the water. The society was duly established; each family had a representative as a member of the society with an equal right to the water. A "coupon" system was introduced and families with little or no land could thus sell their water rights or use the water to share-crop on land belonging to others who were short of water. Any member whose livestock was found grazing in the hills lost his or her rights to the water. The society was given responsibility for maintaining the dams and their catchments, for distributing the water, and for maintaining records.

From then on the village began to develop rapidly. The villagers sold off their goats and replaced them with high-yielding buffaloes to provide milk for the growing towns nearby. The buffaloes are stall fed, using the rapidly growing fodder grasses in the hills. The forest department decided to give the grass-cutting rights to the village rather than, as previously, to a private contractor. They also granted the village the right to grow bhabbar grass in the catchment and machines were introduced to turn the grass into ropes, for which there is a large demand. This, particularly, provided a source of income for the poorer, landless members of the community. Over the past decade the livelihoods of the villagers have continued to grow in diversity and in value, while the hills above are now beginning to be well covered in vegetation and the rates of erosion have dramatically fallen.

Guinope

The second example concerns the Guinope area of Honduras.[22] Like Sukhomajri it was afflicted by severe erosion, compounded by continuous monocropping of maize. The yields were extremely low (around 400 kg/ha). Many farmers were having to make long journeys, even going by bus, to find arable land. There was considerable migration to the slums of the nearby towns; among those who stayed, malnutrition was increasing.

The Guinope project was a collaborative effort between the Ministry of Natural Resources, a private voluntary organization (PVO) in Honduras – the Association for the Co-ordination of Development Resources (ACORDE) – and an American-based PVO, World Neighbours, which had pioneered a low cost, participatory approach to development in neighbouring Guatemala. The project began by identifying the key limiting factor – poor soil quality – and then trying out a small number of appropriate technologies which promised immediate and significant returns. These turned out to be simple on-farm conservation practices, such as contour or drainage ditches, contour rock walls or grass barriers and in-row tillage – most of which had been proven from previous experience in Guatemala – and the use of chicken manure to increase maize yields. These were all tried by the farmers as experiments on their own land. They were encouraged to keep simple accounts and to share results with each other.

These first steps immediately brought benefits – in some cases yield increases of three or four fold – and helped enlist villagers in further experiments in improvement. For instance, the farmers began to experiment with green manures, which had not been tried in the Guatemalan project, and developed new technologies based on velvet bean and lablab bean. As maize yields increased and subsistence was assured, the farmers began to turn to vegetable growing. This entailed further experimentation, the building of a vegetable store and ventures into marketing which initially suffered from considerable set-backs but eventually proved successful. The project is entirely on a self-help basis, with no subsidies provided. By the time of the report some 300 km of erosion works had been constructed by the farmers with their own labour. Over the six years the agricultural programme cost some US $254,000, for project staff, vehicles and office – about US $212 per family for each of the 1,200 families affected.

Outmigration has largely ceased and the landless in the area are benefiting from an increase in the daily wage from US $2.00 to US $2.50 and US $3.00. Many landless have now begun to establish rights to lands which previously they had considered useless, but which under the new technologies are proving productive. Physical and biological sustainability appears to be assured by the emphasis on soil erosion work and the use of indigenous manures. Social and institutional sustainability is being encouraged by the formation of village-level agricultural clubs to co-ordinate and share the results of experiments, a village producers' association which runs the vegetable store, and the training and subsequent employment of villagers as extension agents.

Ingredients of success

The analysis of these and other success stories presented at the conference suggested that there were five key ingredients of success:[23]

(1) *The importance of pursuing an iterative learning approach.* The Sukhomajri project began as a conventional soil con-

servation project but was transformed over time through intensive interaction between the project staff and the villagers into a genuine community project. In Guinope the learning process was built in from the start.

(2) *A conscious decision to put "people's priorities first".* Benefits were clearly apparent and accessible to all, however poor they were to start with. Far too often, soil conservation projects, in particular, have ostensibly been carried out for the benefit of communities, but have rarely satisfied immediate needs and hence have foundered.

(3) *Security of rights and gains for the poor.* As the Sukhomajri project demonstrates these do not have to be based on individual rights to land but can be communal, providing the allocation is equitable and appropriate both to the nature of the resources being conserved and the benefits that can be derived.

(4) *The importance of self-help.* In Sukhomajri, for example, half the cost of land levelling was born by the farmers and each family contributed an equal share of labour for laying pipes, desilting and so on. In Guinope self-help was the corner-stone of the project. The aim was to arrive at a situation in which the villagers only worked because they could see a successful outcome by their own standards. The importance of self-help in the context of sustainability is that it enhances people's capacity to innovate and adapt and so provides them with skills and abilities for the future. In Guinope farmers were encouraged to undertake small-scale experiments, to keep simple records and to share them with each other. Out of the project came a loose-knit federation of village-level agricultural clubs, which shared their knowledge and began to co-ordinate experiments.

(5) *The need for "good" project staff.* Good staff are defined by their sensitivity to farmers' needs, their capacity for insight, and their competence. They also share a capacity for hard work, determination, self-sacrifice and dedication, together with a willingness to stay with projects for a long time and see them through.

Notes

1. There is an extensive literature on FSR&E. Some key recent works are: W.W. Shaner, P.F. Philipp and W.R. Schemehl, *Farming Systems Research and Development Guidelines for Developing Countries* (Boulder: Westview Press, 1982); D.W. Norman, E.B. Simmons and H.M. Hays, *Farming Systems in Nigerian Savanna: Research and strategies for development* (Boulder: Westview Press, 1982); J. Remenyi (ed.), *Agricultural Systems Research for Developing Countries* (Canberra: Australian Council for International Research, 1985); N.W. Simmonds, *Farming Systems Research, a Review* (Washington, DC: World Bank, 1985); "A short review of farming systems research in the tropics", *Experimental Agriculture*, vol. 22 (1986), pp. 1–13.

2. Gordon R. Conway, *Helping Poor Farmers – A Review of Foundation Activities in Farming Systems and Agroecosystems Research and Development* (New York: Ford Foundation, 1987).

3. Michael Collinson, "On-farm research and agricultural research and extension institutes", *Agricultural Administration (Research and Extension) Network*, Discussion Paper 17 (London: Overseas Development Institute, 1987).

4. Gordon R Conway, *Agroecosystem Analysis for Research and Development* (Bangkok: Winrock International, 1987).

5. For discussions of the concept of livelihood in development see World Commission on Environment and Development, *Food 2000: Global policies for sustainable agriculture* (London: Zed Books, 1987); Robert Chambers, *Sustainable Livelihoods, Environment and Development: Putting poor rural people first*, Discussion Paper 240 (Brighton: Institute of Development Studies, University of Sussex, 1987); Robert Chambers, *Sustainable Rural Livelihoods: A strategy for people, environment and development*, Commissioned Study no. 7 (Brighton: Institute of Development Studies, University of Sussex, 1986).

6. D. Werner, N.M. Flowers, M.L. Ritter and D.R. Gross, "Subsistence productivity and hunting effort in native South America", *Human Ecology*, vol. 7, 1979, pp. 303-15.

7. P. Van de Poel and H. Van Dijk, "Household economy and tree growing in upland Central Java", *Agroforestry Systems*, vol. 5 (1987) pp. 169–84.

8. Quoted from Robert Chambers, *Sustainable Livelihoods, Environment and Development: Putting poor rural people first*, op. cit.

9. R. Chambers and M. Leach, *Trees to Meet Contingencies: Savings and security for the rural poor*, Discussion Paper 228 (Brighton: Institute for Development Studies, University of Sussex, 1987); G.F. Murray, "The wood tree as a peasant cash-crop: an anthropological strategy for the domestication of energy", in Charles Foster and Albert Valdman (eds), *Haiti – Today and Tomorrow: an interdisciplinary study* (Lanham, Md.: University Press of America, 1984), pp.141–60; G.F. Murray, "Seeing the forest while planting trees: an anthropological approach to agroforestry in rural Haiti", in D.W. Brinkerhoff and J.C. Garcia Zamor (eds), *Politics, Projects and People: Institutional development in Haiti* (New York: Praeger, 1986), pp.193–226.

10. World Bank, *Indonesia – Java Watersheds*, op.cit., ch.6; G. Feder, "The economic implications of land ownership security in rural Thailand" (draft) (Washington, DC: World Bank, 1986); and David E. Harper and Samir A. El-Swaify, "Sustainable agricultural development in northern Thailand: soil conservation as a component of success in assistance projects", paper presented at the *Workshop on Soil and Water Conservation on Steep Lands* (San Juan, Puerto Rico, March 22–27 1987).

11. For recent reviews of the common-property problem in developing countries see William B. Magrath, *The Challenge of the Commons: Non-exclusive resources and economic development: Theoretical issues*, (Washington, DC: World Resources Institute, 1986); Carlisle F. Runge, "Common property and collective action in economic development", *World Development*, vol. 14, no. 5 (1986), pp.623–35; and Robert Wade, "The management of common property resources: finding a cooperative solution", *Research Observer*, vol. 2, no. 2 (July 1987), pp.219–34.

12. Ian Livingstone, *The Common Property Problem and Pastoral Economic Behaviour*, Discussion Paper 174 (Norwich: School of Development Studies, University of East Anglia, 1985); and David W. Pearce, The sustainable use of natural resource in developing countries, Discussion Paper 86–15 (London: Department of Economics, University College London, 1986).

13. Michael Lipton, *Land Assets and Rural Poverty*, Staff Working Papers no. 774 (Washington, DC: World Bank, 1985).

14. See, for example, Joyce L. Mook, *Understanding Africa's Rural Households and Farming Systems* (Boulder: Westview, 1986); Irene Dankelman and Joan Davidson, *Women and Environment in the Third World* (London: Earthscan Publications, 1988); Susan V.

Poats, Marianne Schmink and Anita Spring, *Gender Issues in Farming Systems Research and Extension* (Boulder: Westview Press, 1988).

15. Pauline E Peters, "Household management in Botswana: cattle, crops and wage labour", in Joyce Mook, op. cit., pp.133–54.

16. There is a growing literature on participatory approaches to rural development. Some of the key works are: Robert Chambers, *Rural Development, Putting the Last First* (Harlow: Longman, 1983); R. Chambers and B.P. Ghildyal, "Agricultural research for resource-poor farmers: the farmer-first-and last model", *Agricultural Administration*, Vol. 20 (1985), pp.1–30; Paul Richards, *Indigenous Agricultural Revolution: Ecology and Food Production in West Africa* (London: Hutchinson, 1985); Paul Richards, *Coping with Hunger: Hazard and experiment in an African rice-farming system* (London: Allen & Unwin, 1986); Ralph Bunch, *Two Ears of Corn: A guide to people-centered agricultural improvement* (Oklahoma: World Neighbours, 1985, 2nd edn); Robert Rhoades and R.H. Booth, "Farmer-back-to-farmer: a model for generating acceptable agricultural technology", *Agricultural Administration*, vol.11 (1982), pp.127–37; John Farrington, "Farmer participatory research: editorial introduction", *Experimental Agriculture*, vol.24, no.3 (1988), pp.269–79 and subsequent six papers in this issue.

17. D.M. Maurya, A. Bottrall and J. Farrington, "Improved livelihoods, genetic diversity and farmer participation: a strategy for rice breeding in rainfed areas of India", *Experimental Agriculture*, vol.24, no.3 (1988), pp.311-20.

18. Paul Richards, *Indigenous Agricultural Revolution* (London: Hutchinson, 1985).

19. Jacqueline A. Ashby, Carlos A. Quiros and Yolanda M. Rivera, "Farmer participation in on-farm trials", *Agricultural Administration (Research and Extension) Network*, Discussion Paper 22 (London: Overseas Development Institute, 1987).

20. See Czech Conroy and Miles Litvinoff (eds), *The Greening of Aid* (London: Earthscan Publications, 1988).

21. P.R. Mishra and Madhu Sarin, "Social security through social fencing – Sukhomajri and Nada's road to self-sustaining development", in Conroy, op. cit; Kamla Chowdry *et al.*, *Hill Resource Development and Community Management: Lessons learnt on microwaters land management of cases of Sukhomajri and Dasholi Gram Swarajya Mandal* (New Delhi: Society for Promotion of Wastelands Development, 1984).

22. Roland Bunch, "Case study of the Guinope Integrated Development Program, Guinope, Honduras", in Conroy, op. cit; and Roland Bunch, *Two Ears of Corn*, op. cit.

23. Robert Chambers, *Sustainable Rural Livelihoods*, op. cit.

6. Conclusions

In the preceding chapters we have concentrated on the critical issues which need to be addressed if a productive, sustainable and equitable agriculture is to be achieved in the developing countries. As we have stressed throughout, progress depends on evaluating and resolving numerous tough trade-offs at all levels of intervention, from international trade down to the individual farm. We do not pretend to have all the answers and, indeed, this book should be seen more as providing a framework for raising issues and setting priorities for analysis and action than giving definite answers. Nevertheless, we have indicated a number of approaches which we feel will, together, make for a programme that promises real progress. In this concluding chapter we summarize our suggestions, indicating what international conditions we need to be *aware* of, what national policies we need to *advocate* and what approaches at the local level we need to *adopt*, to ensure this goal of agricultural sustainability.

The continuing importance of population

It is perhaps an obvious point, but achieving a stable population is an essential precondition for a truly sustainable development. The real achievements of the green revolution have been to maintain and, in some parts of the world, increase the per capita production of food, but it is not at all clear that food production can continue to keep pace with population growth. Even if global production matches global demand, food security may continue to deteriorate because there are growing numbers of people without the land to produce their own food, or the purchasing power to buy it. In India, for example, the proportion of the population below the poverty line persists at about 40%, i.e. over 300 million people, and although the population growth rate is falling (currently 2.1%), the number below the poverty line will exceed 500 million by the year 2025 if present trends continue.

In Africa all the indications are that per capita production will continue to decline. This is not primarily because of growing populations, but it is clear that in countries with particularly high growth rates the decline is likely to be especially severe. In Kenya, for instance, a population growth rate of 4.1% threatens to destroy some very real gains in achieving productive and sustainable systems of agriculture. The tiny 0.2 ha farm in the western provinces of Kenya, depicted in Figure 5.1, is a model of farmer innovation, furnishing what appears to be a highly sustainable although barely sufficient livelihood for the family living there. But the family already consists of a middle- aged couple and three children, one daughter in turn having a child. Because the farm is already highly intensively cultivated, it is difficult to see how its productivity can be dramatically increased. The hope for the family is that the expenditure on education (paid for by the tree cropping) will result in sufficient off-farm income for the children as they grow older. But for this and thousands of other smallholdings in Kenya, the great ingenuity that has gone into creating productive and sustainable agroecosystems will be lost if population growth is not dramatically curtailed.

To a large extent policies to reduce population growth lie firmly in the province of national responsibility. None the less, there is a crucial international dimension, in the provision of support for such policies, and in the provision of aid through multilateral and bilateral funding for the technical and institutional needs of family planning and related health programmes.

International economic relations

Developing countries are increasingly being urged to participate in the global economy, particularly through the export of agricultural commodities. Yet the current unstable international economic climate – with major swings in the dollar, high real interest rates and general financial uncertainties – coupled with chronic indebtedness in many developing countries, is undermining the benefits of such participation.

As small actors in the global economy, developing countries exert little influence on international economic conditions. The control of the major trading currencies, of world interest rates and of financial conditions in general, will continue to rest in the hands of the seven major industrialized nations – the United States, the United Kingdom, Japan, West Germany, France, Canada and Italy. And they, inevitably, will endeavour to create an international financial climate favourable to their own interests. The periodic economic summits of the "Group of Five" and the "Group of Seven", aimed at co-ordinating national economic strategies and a common approach to international economic management, are evidence of how vital the major industrialized countries see the need to exert control over the global economy.

So far they have seen the interests of the developing countries as of peripheral concern. It is true that, in recent years, there has been increasing attention to the threat to global financial stability posed by the Third World debt problem. The so-called "Brady Initiative", launched by the United States Treasury Secretary, has been one attempt to find common agreement on a solution to this problem among the "Group of Seven", the commercial banks, and the most heavily-indebted developing countries. Yet this and other initiatives have resulted less from concern over the economic and social damage inflicted on the developing countries by high levels of debt repayment, than from fears that the inability of certain key debtors to pay back their loans may destabilize the global financial system. With a lessening of these fears, the political will to solve the debt crisis is evaporating, leaving some of the most hard-pressed developing countries with their problems unsolved.

A similar attitude characterizes North–South economic co-operation. In the 1970s the Brandt Report argued that the advanced economies of the North needed to treat developing countries of the South as "equal partners" in global economic relations and development, through a "new international economic order". But to a large extent this plea for greater co-operation has been ignored, essentially because there was little incentive for the advanced economies to create a "new" international economic order when the existing one seemed to be continuing to serve their interests well enough.

Now there is a new international concern – the environment and global sustainability. The Brundtland Report argues that Northern and Southern countries have a "common cause" in the need for more sustainable global development. The industrialized countries are being urged to provide more environmentally-sensitive aid-flows to developing countries – including investments which reduce energy and material consumption – to unilaterally relieve debt burdens, to remove trade barriers to imports from developing countries, and so on. However, as critics were soon to point out, if the industrialized nations have little incentive to encourage greater *co-operation* with the Third World, what incentive do they have to make actual *concessions* in order to encourage more sustainable global development?

Nevertheless, it is becoming apparent that public concern about the environment and global sustainability continues to grow in both the North and the South. The greenhouse effect, ozone-layer depletion, deforestation and land degradation are now major political issues in all countries. While the environmental concerns of the 1960s and 1970s were largely local in their impact, these new concerns are daunting in the scale of their potential effect and in the corresponding magnitude of the policy responses required. The effects of global warming on agriculture, for instance, are likely to be widespread and, for some countries, catastrophic. Increasing global temperatures imply shifting climatic zones with, in many regions, much greater climatic variability. Certain countries will suffer more from droughts and floods.

At greatest risk in the developing countries are:

● the lowland areas and island countries of the humid tropics in Asia, the Pacific and the Caribbean

- the arid and semi-arid tropics of Africa and South Asia, and the Mediterranean climate of West Asia and North Africa
- the rainfed uplands and highland regions, particularly with poor soil conditions.[1]

Such is the present level of concern over global warming that there have been serious proposals for such measures as a "carbon tax" in industrialized countries to reduce the emissions of carbon dioxide. The evidence for ozone-layer depletion has already led to the Montreal Protocol, committing all signatories to reduce and eventually eliminate production of chlorofluorocarbons (CFCs). Serious global negotiations are also proceeding over international compensation, including debt relief, to encourage tropical countries to reduce the rate of deforestation and to preserve areas of unique biological diversity. And virtually all bilateral and multilateral donor and lending agencies are reviewing their programmes and aid strategies in order to improve their ability to assist more sustainable management of natural resources.

In all these situations the industrialized countries are becoming aware that they cannot solve problems unilaterally. Emissions of carbon dioxide and CFCs have to be reduced globally. Even if per capita pollution is much higher in the industrialized countries, the developing countries are significant and growing polluters. Mutual agreements are essential. There is thus now a real possibility of North–South co-operation over global *environmental* issues. Ironically this may be the catalyst for bridging the seemingly intractable divisions between industrialized and developing countries over global *economic* issues.

Commodity Prices
One such issue needing urgent attention is the long-term decline in real prices of globally-traded commodities. Very little can be done to reverse those trends which have determined the weakness of demand for raw materials in industrialized countries. Substitutions in consumption, or changes in tastes, or the development of production processes which use raw materials less intensively, are likely to be permanent. It is true that overproduction of certain export crops in developing countries,

due to price rises of the 1970s, seems to be petering out, but many countries are still trying to boost production so as to increase export revenues or their share of the world market. Perhaps the most significant measure which could be undertaken is a reduction in the subsidies to, and protection of, domestic agriculture in industrialized countries.

As we argued in Chapter 3, present agricultural trade and pricing policies in developing and developed countries are rarely either efficient or sustainable. From the developing countries' viewpoint, increasing agricultural production of exportable commodities is undesirable if the associated total costs – which include the financial burden of subsidies, the cost of environmental impacts and the social costs of regional changes in cropping patterns, farming systems and crop output (e.g. from "food" to "export" crops) – exceed the benefits of increased export earnings or share of the world market. Too often, only these benefits are considered while the costs are ignored or overlooked. But it needs to be realized that this is no longer a realistic strategy, given the long-term trends in real prices, the frequent scarcity of domestic financial resources and the evidence that sustainable agricultural development depends on a careful management of the natural resource base.

However, such a reform of agricultural policies in developing countries will be largely fruitless without similar steps in industrialized economies. As we have seen, depressed commodity prices are also a result of the domestic agricultural policies and protectionism of the United States and the European Community. This argument, though, still has little force. The effects of these policies on global agriculture, and in particular on production in developing countries, are largely peripheral to the policy debates in industrialized countries. Indeed, it is the "fear" of losing domestic markets to external competitors, including developing country producers, that has kept up the pressure for agricultural protectionism in many industrialized economies. If reform does occur, it will be a consequence of the perceived financial burden imposed by these policies on industrialized countries themselves – coupled, perhaps, with concern over any environmental degradation resulting from the subsequent distortions in agricultural production.

Free trade and structural reform
At the end of Chapter 3 we discussed briefly the trade-off between the pursuit of global economic efficiency, through the promotion of free trade and structural reforms in the developing countries, and the goal of increasing sustainability and equity. Our verdict was that without detailed analysis of how national policies affect sustainability and equity issues within their national boundaries it is difficult to come to general conclusions. As Chapter 4 revealed, however, much depends on the nature of the environment, the resources being exploited and the commodity being produced. Trade liberalization may encourage the development of highly sustainable and equitable agroecosystems, yet equally it can provide a licence for highly exploitative systems which are environmentally and socially destructive in the long term.

National strategies

In Chapter 4 we discussed two key strategic issues that illustrate this point. One of the structural adjustment reforms being urged on developing countries consists of a reorientation toward production of agricultural commodities for export. It is claimed that this strategy will help countries reduce their debt burden in ways that make optimal use of available labour. The counter argument is that this is likely to result in serious environmental damage. But, as we pointed out, this objection is too simplistic. There is, first of all, no clear ecological distinction between export and food crops – some food crops, such as rice in Thailand, constitute major exports. The question as to whether promotion of export crops is environmentally sustainable depends on the nature of the crop and the conditions under which it is grown. We gave examples of situations where expansion of export crops in fertile lands is pushing food production into more marginal environments and, on the other hand, where growing of food crops such as maize is replacing the growing of tree crops on hilly land, with consequent increases in soil erosion.

Similar considerations apply to the argument over whether to invest in more favoured or marginal lands. It is often claimed that the best use of agricultural investment, whether it be in research, infrastructure or development projects, is in environments which are well endowed in terms of soil, water, topography and access to markets. By contrast, investment in marginal lands is more problematic and less likely to be sustainable. As we argued, these polarized positions are again too simplistic. There is good evidence that the incremental returns to investment in the best favoured lands is declining and that such returns are for the most part reliant on increasing inputs, such as pesticides and fertilizers, which in both economic and ecological terms are unlikely to be sustainable. On the other hand, although the potential productivity on marginal lands is less than on the best favoured lands, the incremental returns may well be higher. There is a much greater gap between actual and potential productivity in marginal lands than on the best favoured lands. As the two examples of development projects cited in Chapter 5 – Sukhomajri and Guinope – clearly show, it is possible to develop productive, equitable and sustainable systems under the most difficult conditions.

We also mentioned two other key issues of national strategy – investment in small-scale farming versus large-scale farming, and the role of private, versus public, investment. Again, for the criteria of sustainability and equitability much depends on the nature of the agroecosystem under consideration, its socio-economic as well as ecological features. Some large-scale agriculture, for instance rubber plantations in Malaysia, has proven highly productive and sustainable. On the other hand, small-scale "pioneer" shifting cultivators in the forests of Indonesia and Brazil are some of the most environmentally destructive farmers in the world. Similarly, while it is theoretically easier for public agricultural investment to take a longer term view and build in sustainability practices, the record shows that this does not necessarily happen. There are numerous examples of publicly funded agricultural developments which have been disastrous environmentally – the notorious "Groundnuts Scheme" in Tanzania in the 1950s, for instance. Even the highly successful and apparently sustainable Gezira cotton scheme in the Sudan has suffered severe declines in productivity in recent years due to reduction in the fallow period, too high an intensity of cropping and serious pest problems which appear, in part, to be due to pesticide overuse. By contrast, there are an equal number of examples of sustainable private investments. One such is the little known expansion of Turkish tobacco in north-east Thailand which has brought considerable economic benefits to small farmers in a highly impoverished area, and in such a manner that cultivation exploits, without destroying, the potentials of very marginal environmental conditions.

Policy criteria

What this discussion highlights is that such controversial issues cannot be resolved in isolation. Whether sustainable and equitable agricultural development is attained or not depends on the broader policy environment and the degree to which it is in tune with the biophysical environment of agriculture.

We further suggest that the creation of such a favourable policy environment depends, in turn, on whether developing nations are able to meet five crucial criteria:

(1) Governments will only be concerned with long-term natural resource management issues, especially the appropriate development of resource-poor lands, if they perceive it to be economically essential to do so. *We refer to this as the "political will" criterion;*

(2) Proper economic analysis of policy options, especially of their impacts on small and marginal farmers in resource-poor regions, requires appropriate data, methodology and analytical tools for economic valuation of environmental impacts. *This is the "economic analysis" criterion;*

(3) Small farmers and pastoralists will only change their farming practices and current uses of the resource base if they have the appropriate economic incentives to do so. Equally, their response to programmes and projects is only likely to be sustainable if they are genuinely involved in their design and implementation. *This is the "appropriate incentives" criterion;*

(4) Policies and programmes for sustainable development will only be properly implemented if the appropriate institutional framework is established (e.g. one that is multisectoral, coordinated over natural resource boundaries, and "bottom-up" in orientation). *This is the "institutional flexibility" criterion; and*

(5) Sustainable agricultural development cannot be pursued in isolation. Its success depends on its place within an infrastructural development which complements and reinforces sustainable policies and programmes. *This is the "complementary infrastructure" criterion.*

The political will

The "political will" criterion can be met in a number of ways. One is for international lending agencies and other donors to insist, as part of the conditions of structural adjustment, that governments of developing countries adopt sustainable agricultural policies. Second, governments may be persuaded that sustainable management of their natural resource base is essential to meeting their debt obligations and long-term development goals. Third, they may become convinced of the significance of the potential economic contribution of smallholders and pastoralists on marginal lands. These, however, are interlinked arguments. The acceptability of "conditionality" will depend, to a large extent, on the acceptance of the last two arguments. It is unlikely that "conditionality" on its own will be sufficient in meeting the "political will" criterion.

But there is, in our view, a much more powerful argument which we advanced in Chapter 3; a large number of low and lower-middle income economies are directly dependent on agricultural commodities for the overwhelming majority of their exports, and this is likely to remain true for a long time to come. In this situation export performance is a direct function of the efficient and sustainable use of the natural resource base that supports agricultural production. Furthermore, in the absence of sustainable resource use policies such countries will become increasingly vulnerable to the economic stresses imposed by external debt.

Economic analysis

However good these arguments in theory, they have to be backed up by detailed economic analyses both at the macro and micro level. For instance, there is a need for substantive and extensive analysis of the implications of various macroeconomic, trade and sectoral policies on the local resources upon which the livelihoods of smallholders and pastoralists depend. In particular, better evaluation of the economic costs of these environmental impacts is required. At the micro level, attention has to be paid to the natural resource allocation decisions made by farmers and village communities, both to help design appropriate policies and investment programmes and to monitor their impacts.

There are two problems facing analysis of this kind. The first is the lack of a data base and a methodology for evaluating resource and environmental impacts.[2] Current data bases in developing countries, where they are reliable, are disaggregated by administrative and political boundaries (i.e. region, province, district, sub-district and so on). It is often extremely difficult to obtain the same economic and environmental data by major agro-ecological and resource system zones; e.g. watersheds, semi-arid lands, uplands, forests, coastal resource systems. Equally, it is difficult to obtain reliable data on certain key marginal socio-economic groups, such as agropastoralists, nomads, upland farmers, shifting cultivators and indigenous tribes. The second problem is that although techniques for evaluating the environmental impacts of economic policies and projects have been developed in recent years, they have yet to be applied in developing countries.[3] Systems of natural resource accounts, which attempt to show both the full economic contributions of the resource base and the costs of its depletion and degradation, have been greeted with enthusiasm in some developing countries, but still have to be systematically adopted.[4]

Appropriate incentives

Meeting the first two criteria establishes the need and the will to act. The success of that action depends on the remaining three criteria. The first of these is the capacity to design "appropriate incentives", of which two kinds can be distinguished.

"VARIABLE" INCENTIVES

These focus on price changes to induce producers and consumers to manage natural resources in a more sustainable manner.[5] They include altering input and output pricing, exchange-rate modification, tax and subsidy reform, adjusting middlemen margins, and so on.

In theory such policies should provide economic incentives for smallholders and pastoralists to increase productivity without generating environmental degradation. However, in practice it is not quite so simple. As the example of rising food prices discussed in Chapter 4 illustrates, different rural groups will be affected differently by changing prices. Moreover, there is still little empirical understanding of the linkage between price changes and agricultural supply and demand responses, and natural resource effects.[6]

In the case of subsidies of agricultural inputs, a more substantial link has been established between wasteful and inappropriate use of these inputs and environmental pollution and degradation.[7] But we are far from being able to determine the optimal level of fertilizer, pesticide and other input use in terms of sustaining agricultural production in developing regions. Thus in some countries, like Nepal, natural geographical constraints may require the continuation of transportation subsidies for fertilizers and other inputs in order to ensure distribution to remote areas. But other factors have to be taken into account. Nepal is not free to set its own input subsidies but must keep parity with its dominant neighbour, India. If Nepal's fertilizer prices fall too low, scarce supplies are rapidly smuggled across the large open border.[8] The optimum level of subsidy is thus, at present, largely an empirical compromise.

Moreover, working with only one set of these incentives is likely to be ineffective. The challenge for agricultural policy is to design the right combination for a given target group of farmers and pastoralists.

"USER ENABLING" INCENTIVES

These directly address the needs of the resource user. They include changes in land and resource rights, for example, and, most important, increased participation in decision making. Their design, inevitably, requires careful analysis of the impact of institutional arrangements for land ownership, and rights of access and tenancy, on environmental degradation. Such analysis, though, is complicated by the fact that few Third World rural households are bound by one set of institutional property rights; many own little land, rent in a little more, do some farm labour for other, bigger owners and even have some rights over certain commonly-owned resources. Moreover, the relationships between resource access, especially to land, and poverty may be completely different in marginal as opposed to irrigated agricultural regions with more favourable environmental conditions. Under irrigated, improved or modern farming conditions that produce high net returns per hectare, access to even a little bit of land, despite being associated with larger household size, tends to reduce the probability of poverty in an average year. On very resource-poor lands, however, farmers with small and even middle-sized holdings tend to be only marginally better off than landless labourers.[9]

The design of appropriate user-enabling incentives also requires a good understanding of how economic incentives determine the behaviour of farmers and pastoralists. Profitability is often a powerful motive, even for resource-poor rural households. The studies of upland farming on Java, referred to in earlier chapters, suggest that (a): farmers will modify their land management practices and farming systems to improve soil and water conservation if they perceive an economic advantage from doing so and (b); this "economic advantage" is largely determined by the potential for increasing productivity and thus net returns from working the land. This decision will be affected by such factors as the ability to earn greater returns from off-farm employment, the security of land tenure, transportation and marketing facilities and the access to information on technology, inputs and farming methods. But, in general, the relationship between the erodibility and profitability of different farming systems on different soils and slopes is a critical determinant of whether upland farmers adopt a soil conservation strategy.[10]

Incentives to greater participation by small farmers in development programmes and projects are largely a function of creating the right institutional framework, and this is addressed in the next section.

Institutional flexibility
In many ways this is the most difficult of the criteria to meet. It is not simply a question of central governments devolving power automatically to provincial or regional authorities or below. There may be formidable political obstacles to this approach and a lack of institutional and administrative competence at the decentralized levels. One answer, in the short term, is to establish stronger sectoral and sub-sectoral links among existing ministries committed to improving the integration and co-ordination of the activities in terms of new sustainability targets. In the past there has been too great a reliance on single physical planning targets, such as increases in yields of specific crops, the number of trees planted or the number of dams built, as the measures of performance. More attention needs to be paid to improving the technical and administrative capacity across all agencies for managing activities based on natural resource system boundaries and zones (e.g. watersheds, coastal zones) rather than political/administrative units, and on improving the performance of farming systems and livelihoods rather than specific commodity crops.

Among other things, this implies incentives for rural extension workers to work effectively in the remote regions where resource-poor farmers and pastoralists are located. Development institutions at the local level have to be strengthened and given a new sense of purpose and ideology, based on a "bottom-up" approach to development. Techniques such as Rapid Rural Appraisal now exist to make such an approach a practical rather than a theoretical proposition. In the hands of both research and extension workers they have proven successful both in ensuring local participation and in providing appropriate analysis for sustainable agricultural development (see Appendix).

It needs also to be recognized that such invigoration of local development competence cannot be left solely in the hands of government. In recent years the pioneers of "bottom-up" approaches have frequently been non-government organizations (NGOs), imbued with a sense of idealism and unconstrained by bureaucratic procedures. In particular they have been innovators in the development and implementation of sustainable technologies and village-level institutions. Their role needs to be encouraged and supported, particularly where, as in the case of Sukhomajri and Guinope, they operate in close partnership with local government agencies.

Complementary infrastructure
Finally there is a need to encourage, particularly in marginal agricultural areas, co-ordinated rural development efforts which combine economic incentives with appropriate physical infrastructure and institutional investments. These will include improved marketing and transport, post-harvest technology and processing, the provision of rural credit, and the development of complementary research and extension.

Besides their contribution to agricultural production, such infrastructure investments in rural areas also have the potential for direct and indirect generation of off-farm employment. Thus the increased investment in rural infrastructure, which accompanied the rice-based expansion in the lowlands of Java, resulted in expanded off-farm employment opportunities in trade, transportation, private construction and services, and this especially benefited the landless and those with marginal holdings. Greater infrastructure investments in hitherto neglected marginal areas could have similar important income-generating and multiplier employment effects. One example is the establishment of food processing industries in rural areas which generates the need for storage, transportation, sorting, grading and packing in addition to the actual processing.[11]

There is some evidence that the availability of off-farm income may lessen farmers' attachment to the land and hence their willingness to invest in improved land management. But if physical infrastructure investment is well co-ordinated with agricultural and rural development activities, the effect should be to expand overall incomes and employment opportunities so that the additional resources will be invested in land improvements.[12] This is, after all, one of the little-acknowledged lessons learned from the co-ordination of investment efforts that established the green revolution on the lowland rice-growing areas of Java.

Similarly, investment in agricultural research and extension needs to be redirected away from the more favoured agricultural regions and their crops. In Thailand, until recently, the irrigated rice-growing Central Plains received most of agricultural investment, whereas the mainly rainfed, more marginal, north-east region received little over 10% of the official development assistance for agriculture.[13] In Indonesia, where expenditure on agricultural research is far below that of Thailand and comparable economies, amounting to only 0.3% of agricultural GDP, the overwhelming emphasis is on research on rice that favours the irrigated lowland areas.

In Sub-Saharan Africa the situation is even worse. The needs of small countries and the diversity of agroclimatic conditions were matched by a mere US $170 million of agricultural research expenditure in 1980. The World Bank recommends this is doubled in real terms by 1990. But perhaps more important is the nature of the research investment. Much current research fails to meet the region's needs. This is linked to two causes: an inadequate understanding of small-farmer goals and resource limitations – for example the vast importance of intercropping compared to monocrop systems; and (with a few exceptions, such as hybrid maize in Zimbabwe and Kenya and irrigated rice where environments can be modified to suit the crop), the inappropriateness of transferring green revolution technologies which were successful in Latin America and Asia. Africa has a higher rate of demographic change, comparatively low-input agriculture and more difficult agroclimatic conditions.[14]

Sustainable livelihood development

The issues of appropriate incentives, infrastructure and institutions lead directly to a consideration of the tasks facing sustainable and equitable development at the local level. Working with only one set of incentives is unlikely to be effective. The challenge for policy makers is to design the appropriate combination of incentives for particular groups of farmers in particular, well-defined and understood environmental and socio-economic circumstances. Appropriate infrastructure investments and institutions must also take these circumstances into consideration. These are necessary conditions for success, but by themselves are not sufficient.

In the last chapter we described two examples of successful development projects that hold the promise of sustainability. We listed some of the elements of their apparent success – iterative learning, security of rights, a bottom-up, self-help approach and the presence of dedicated development staff. But there is also a more fundamental ingredient: the process of seeking out and agreeing on a series of "deals" which minimize the trade-offs between productivity, stability, sustainability and equitability.

The Sukhomajri case study makes this very clear. Success there has depended on a series of deals or agreements between the government agencies and the villagers, between the poor and landless and the better-off in the village, between the milk producers and the middlemen and so on. In these deals everyone has benefited to a considerable extent. The government has achieved its goals of arresting watershed erosion and reforestation, while the village has grown in prosperity. Moreover the increased wealth has been relatively equally shared. Most important, stability and sustainability have been assured not only because of the nature of the deals but also because they have been freely entered into and have become part of the new institutional arrangements of the locality.

Perhaps there are broader lessons here. In this book we have stressed three central themes of sustainable agricultural development:

- that incorporating sustainability of agricultural production as a development objective requires explicit recognition and

understanding of the trade-offs involved with other objectives
- that the uniqueness of each production system in the agricultural hierarchy and the hierarchical linkage between the different levels means that the problem confronting sustainable agriculture must be tackled at all levels – local, national and international
- that proper analysis of sustainable agriculture for development requires a consideration of the trade-off between sustainability and other development objectives among, as well as within, the different levels of the agricultural hierarchy.

What we are suggesting is that, once such analyses are carried out, the route to sustainable and equitable agricultural development depends on seeking out and implementing those agreements, both within and between the levels in the agricultural hierarchy, that will minimize the trade-offs and maximize the manifold benefits from achieving high levels of productivity, stability, sustainability and equitability.

We do not suggest that we have found the ultimate solution to the questions we have raised. But we do believe that this book provides a comprehensive framework for thinking about sustainable agriculture for development – a framework that has been built upon the simple but non-trivial notions of hierarchical linkage and trade-off.

Notes

1. Edward B. Barbier, "The global greenhouse effect: economic impacts and policy considerations", *Natural Resources Forum* (February 1989), pp.20–32; and P.A. Oram, "Sensitivity of agricultural production to climatic change", *Climatic Change*, vol.7 (1985), pp. 129–52.
2. Edward B. Barbier, "Economic valuation of environmental impacts", *Project Appraisal*, vol.3 (1988), pp.143–50; and David W. Pearce, Edward B. Barbier and Anil Markandya, *Environmental Economics and Decision-making in Sub-Saharan Africa* (London: LEEC Paper 88-01, 1988).
3. Some of these valuation techniques are described in John A. Dixon, Richard A. Carpenter, Louise A. Fallon, Paul B. Sherman

and Supachit Manopimoke, *Economic Analysis of the Environmental Impacts of Development Projects* (London: Earthscan Publications, 1988); and Maynard M. Hufschmidt, David E. James, A.D. Meister, Blair T. Bower and John A. Dixon, *Environment, Natural Systems and Development: An economic valuation guide* (Baltimore: Johns Hopkins University Press, 1983).

4. See Robert Repetto, Michael Wells, Christine Beer and Fabrizio Rossini, *Natural Resource Accounting for Indonesia* (Washington, DC: World Resources Institute, 1987); and Yusuf J. Ahmad, Salah El Serafy and Ernst Lutz (eds) *Environmental and Resource Accounting and their Relevance to the Measurement of Sustainable Development* (Washington, DC: World Bank, 1988).

5. See, for example, Pearce, Barbier and Markandya, op cit.; Edward B. Barbier, Anil Markandya and David W. Pearce, *Sustainable Development – Economics and Environment in the Third World* (London: Edward Elgar, 1990); Robert Repetto, *Economic Policy Reform for Natural Resource Conservation*, Working Paper no.4, (Washington, DC: Environment Department, World Bank, 1988); and Jeremy J. Warford, *Environment, Growth and Development*, Development Committee Papers, no. 14 (Washington, DC: World Bank, 1987).

6. See the discussion in Pearce, Barbier and Markandya, op.cit.

7. See Repetto, op.cit.

8. Michael B. Wallace, "Fertiliser Price Policy in Nepal", *Strengthening Institutional Capacity in the Food and Agricultural Sector in Nepal*, Research and Planning Paper Series no.6, HMG-USAID-GTZ-IDRC-Winrock Project, November 1986.

9. Michael Lipton, *Land Assets and Rural Poverty*, World Bank Staff Working Papers, No. 774 (Washington, DC: World Bank, 1985).

10. See Edward B. Barbier, *The Economics of Farm-level Adoption of Soil Conservation Measures in the Uplands of Java*, Working Paper no.11, (Washington, DC: Environment Department, World Bank, 1988); and Brian Carson, with the assistance of the East Java KEPAS Working Group, *A Comparison of Soil Conservation Strategies in Four Agroecological Zones in the Upland of East Java* (Malang: KEPAS, 1987).

11. See, in particular, William L. Collier, Soentoro, Gunawan Wiradi, Effendi Pasandaran, Kabul Santoso and Joseph F. Stepanek, "Acceleration of rural development on Java", *Bulletin of Indonesia Economic Studies*, vol.18, no.3 (1982); and Douglas D. Hedley, "Diversification: concepts and directions in Indonesian agricultural

policy", *Workshop on Soybean Research and Development in Indonesia* (Bogor: The CGPRI Centre, 24-26 February 1987).

12. World Bank, *Indonesia – Java Uplands*, op.cit.

13. Gordon R. Conway, *Sustainable Agroecosystem Development for Thailand*, a Report for USAID Bangkok (London: IIED, 1987); and World Bank, *Indonesia – Agricultural Policy: Issues and options* (Washington, DC: World Bank, 1987), vol.1.

14. Dunstan S.C. Spencer, "Agricultural research: lessons of the past, strategies for the future", in Robert J. Berg and Jennifer Seymour Whitaker (eds), *Strategies for African Development* (Berkeley: University of California Press, 1986) pp.215-41. J. Boesen *et al.*, *Research on the Increasing Vulnerability of Peasant Farming Systems, Their Resource Base, and Food Production in East Africa* (Copenhagen: Centre for Development Research, August 1987), and World Bank, *Toward Sustained Development in Sub-Saharan Africa* (Washington, DC: World Bank, 1984), pp.31–2.

Appendix
Agroecosystem Analysis

Much of this book has been concerned with the difficulties of developing appropriate policies and programmes for sustainable agriculture, but there are also serious practical challenges facing the designers and implementers of sustainable projects. First, since the various components of sustainability – ecological, economic, social and institutional – are so closely intertwined, successful projects must necessarily involve a wide variety of skills and disciplines. As experience shows, getting these to work together in design and implementation is by no means easy. Second, while the sustainability of development is, by definition, impossible to judge in the short term, there are few situations where it is practical to wait for the results of long-term research. Actions have to be taken quickly and efficiently, and as cheaply as possible, using whatever knowledge is to hand. Fortunately, a number of methods and techniques now exist which offer partial solutions to these challenges.

Development of the approach

One such method is that of Agroecosystem Analysis (AEA).[1] Work on the development of this method began at the University of Chiang Mai in northern Thailand about ten years ago. In 1968 a Ford Foundation grant had been given to create a multiple cropping project (MCP) aimed at designing advanced triple-crop, rotational systems which farmers could use to capitalize on the government irrigation schemes which had recently been installed in the Chiang Mai valley. At the same time, many of the young staff were given scholarships to go abroad for further graduate training. In the late 1970s they returned, eager to use their new skills and experience in the task of helping the farmers of the valley. But, almost immediately, they realized that much of the work of the MCP in the intervening years had not proved particularly relevant. Although the MCP had developed some half dozen apparently superior and productive crop systems, there were very few cases of adoption by the farmers; on the other hand the farmers themselves had developed a large number of triple-crop systems in response to the new opportunities that the irrigation had provided. This realization raised questions in the minds of the staff as to the role they, as university researchers, could most effectively play. In terms of helping the farmers of the valley, where did their comparative advantage lie? Should they continue to design new systems? If not, what kind of research should they undertake? They further realized that these questions could not be answered until they had a better idea of the existing farming systems in the valley and the particular problems the farmers were currently facing. The group at Chiang Mai, numbering some twenty academic staff, then spent approximately a year developing an approach that would give them the answers.[2]

The group soon realized that multidisciplinary analysis requires more than simply having a research or development team that works well together and is sensitive to the requirements of good communication. The generation of good interdisciplinary insights also requires organizing concepts and relatively formal, i.e. semi-structured, working procedures. The key concepts they developed – the agroecosystem, agroecosystem properties and hierarchies – have been described earlier in Chapter 2; they are relatively simple and generally acceptable to all disciplines. They are also understandable, at least in essence, by those with whom the development professionals are working; that is, both policy makers and the farmers.

The next step was to use these concepts as a basis for analysis, both in the field and in a workshop environment. It was soon found that the most powerful analytical tools were simple, but well designed, descriptive diagrams. These were prepared in the field, from direct observation and through interviews with farmers. They were then used in a workshop to facilitate communication between the different disciplines and to pin-point the critical problems and opportunities facing farmers and hence to identify the key research priorities. The priorities were laid out as a set of key research questions with accompanying hypotheses and an outline of the research work that was needed. In the years that followed the Chiang Mai team has used the list as a guide to its research and has been able to provide an impressive number of answers to the questions.

The method was taken subsequently to Khon Kaen University in the north-east of Thailand, where it was adapted to the problems of analysing the semi-arid agroecosystems of north-east Thailand,[3] and thence to Indonesia where it was applied to the analysis of the research needs of, respectively, the uplands of East Java, the tidal swamplands of Kalimantan and the semi-arid drylands of Timor.[4] More recently, AEA has been used as a method for determining development priorities for the Aga Khan Rural Support Programme in the northern areas of Pakistan[5] and for the Ethiopian Red Cross Society in Wollo province in Ethiopia.[6] In both these situations the need was for even more rapid multidisciplinary diagnosis and the method has undergone considerable simplification.

AEA is similar to the CIMMYT OFR/FSP (On-Farm Research with a Farming Systems Perspective – see Chapter 5) but differs in several important respects:

(1) an emphasis on the use of multidisciplinary workshops and rapid appraisal techniques;
(2) a foundation on ecological as well as socio-economic concepts;
(3) a recognition of the importance of the trade-offs in agricultural development between productivity, stability, sustainability and equitability; *and*
(4) its applicability not only to farming systems but to the analysis and development of larger systems at the village, watershed, regional and even national level.

The first step in AEA is to carry out some kind of zoning of the project area.

Rapid Agroecosystem Zoning

In most developing countries, agricultural investment and planning is usually channelled down a hierarchy of administrative/political units – from the national and regional level to provincial, districts and sub-districts and finally to the village level. Agricultural data are collected and aggregated through this hierarchy, which serves as the basis for research and extension activity and the dissemination of agricultural improvements. On the other hand, many environmental and resource problems in agriculture do not neatly conform to such administrative or political units, but instead are contained by ecological and socio-economic boundaries. Soil erosion, for example, relates to the extent of an upper watershed, desertification to the range of a pastoral tribe and salinization to an irrigation command area. Moreover, under diverse agroclimatic and socio-economic conditions, even a relatively small-scale administrative/political unit (e.g. a district or sub-district or even village) may contain many diverse and complex resource systems, each with its distinct set of cropping patterns, soil type, institutional arrangements and economic circumstances.

Various methods of land-use appraisal are currently available.[7] They include land suitability analysis, agroecological zoning and life zone classification. All of these are powerful analytical approaches but they are also data hungry, requiring extensive field surveys and detailed information on climate, soils, vegetation and so on. They also concentrate on biophysical to the exclusion of socio-economic variables.

Often, development projects cannot wait for such detailed land suitability analysis. Yet the immediate need is for a characterization of the area into zones such that development innovations tested in one part of a zone should be extendible to other parts of the zone, and possibly to other similar zones in different valleys. Zones defined in this way are referred to as General Recommendation Domains. In development terms they may be regarded as broadly homogeneous. Such zones need not be immutably fixed, however; they may develop incrementally and iteratively, the boundaries changing with time, following the acquisition of more data and greater experience from development interventions.

Most important, General Recommendation Domains need not be characterized solely in terms of biophysical features, but can also be determined by socio-economic factors. In effect this means that the domains are agroecosystems, as defined in Chapter 2, which lie between the village and valley agroecosystems in the hierarchy. The method is accordingly referred to as Rapid Agroecosystem Zoning (RAZ) and is characterized by a rapid, iterative classification process.

The Hunza Valley
An example of such a zoning was carried out in the Hunza Valley of northern Pakistan for the Aga Khan Rural Support Programme.[8] An initial secondary data survey suggested that the primary biophysical determinants of the recommendation domains were likely to be growing period and resource availability, with topography and soils playing an important role within the domains and within the village agroecosystems.

The zoning was begun with growing period since some relevant meteorological data were available. Although temperature data were only available for three locations in the valley, these were supplemented by data from a further three locations in the district to determine a temperature lapse rate for each 500 feet of altitude. From these figures it was then possible to derive growing period as a function of altitude along the valley, taking 5°C as the temperature below which growth ceases.

At this stage a rapid field survey was undertaken in which farmers along the valley were interviewed to determine sowing and harvesting dates, the dates of the first frosts, and so on. The farmers' own sense of zoning was also explored. In each village they were asked to compare their village with neighbouring villages and to point out the major differences – not just in terms of crops or livestock but any feature that occurred to them.

Based on the secondary data and the survey, the valley was zoned as in Figure A.1, and the team produced for each zone a brief portrait and an initial set of strategies for development (Table A.1).

Figure A.1: The nine valley zones of the Hunza Valley produced by the first iteration of the Rapid Agroecosystem Zoning

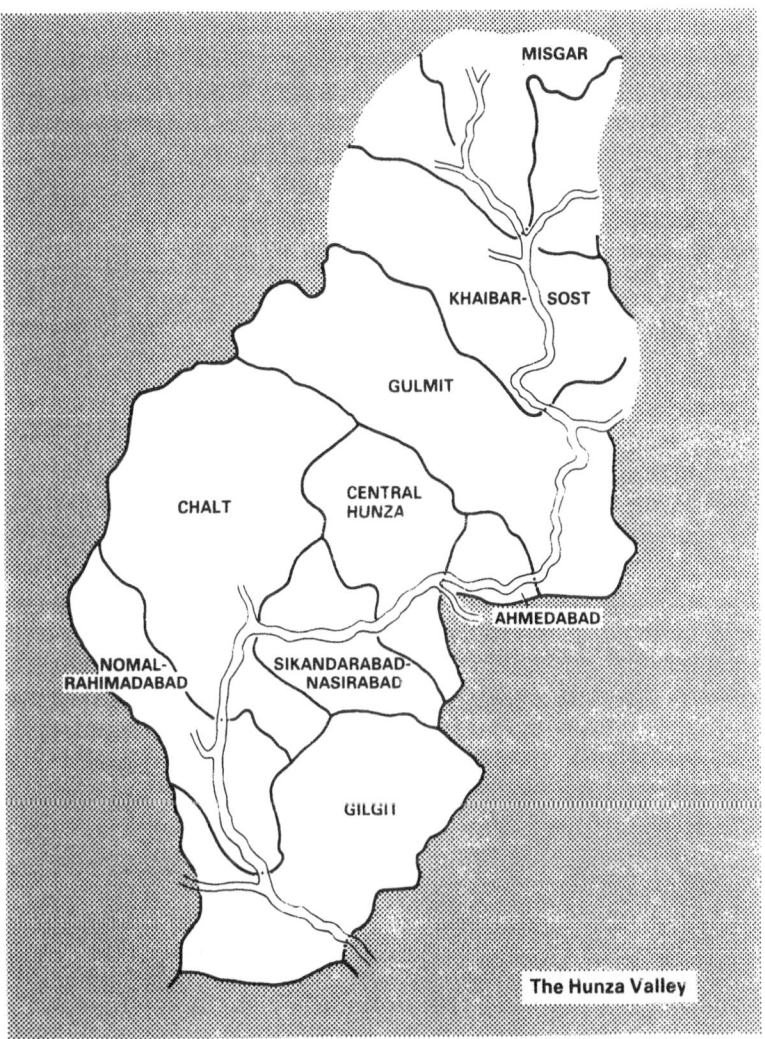

Table A.1: Characterization of the Zones of the Hunza Valley following the first iteration of the Rapid Agroecosystem Zoning

	Zone	Growing days	Crops/yr	Portrait
1.	Gilgit	330	2	Dominated by Gilgit which provides market opportunities for high-value specialist crops, livestock products, timber and firewood that can be produced in the long growing period
2.	Nomal-Rahimabad	310	2	A two-crop zone with a long growing period and good resources which requires a balanced and flexible development of fruit, vegetable, dairying and poultry for the Gilgit market
3.	Chalt	280	2	The highest fully two-crop zone with extensive natural resources. Development should focus on irrigated and natural forest and pasture, with specialization in dairying
4.	Sikandarabad-Nasrabad	280	1/2	A relatively narrow transitional zone with limited land. The major requirement is for early maturing, high yielding wheat varieties and intensification of fruit production, processing and marketing
5.	Central Hunza	260	1/2	A broad bowl of good quality land, all under cultivation, where the future lies in urbanization and intensification of arable land use, emphasising fruit, vegetables, dairying and poultry
6.	Ahmedabad	260	1/2	A narrow, extremely land-scarce zone which is the uppermost area for doubling cropping. The only future lies in land intensification based on new varieties, better cultivation practices and plant protection and a shift to fruit and forest trees
7.	Gulmit	240	1	The lowest of the single-crop zones where the future lies in servicing tourism through provision of high quality crop and livestock products and the immediate priority is stabilization of seed potato production and marketing
8.	Khaiber-Sost	220	1	With a growing period of only 220 days, the future lies in forest production primarily for timber and firewood, intensive livestock production using irrigated fodder, and seed potato production
9.	Misgar	190	1	The uppermost zone in the valley characterized by a very short growing period and a serious shortage of labour. The immediate need is for rapidly maturing cereals and vegetables, and for the growth of seed potato production, while long-term development should emphasize forest production

Figure A.2: Transect of a village in the Hunza Valley

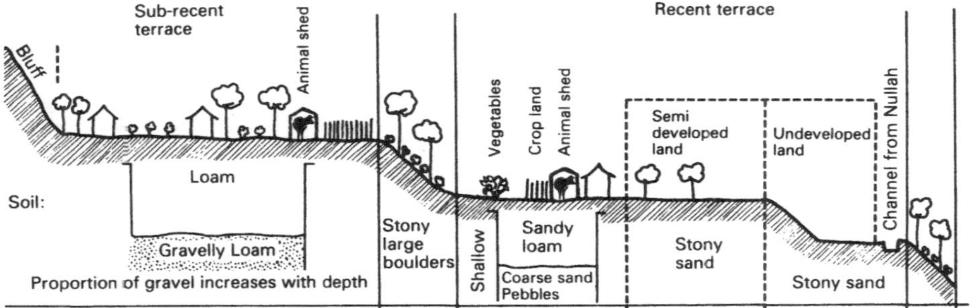

Crops:	Wheat – maize, mash Wheat – fallow Vegetables Potatoes	Bushes Trees	As for recent terrace	Willow Poplar Russian olive Thorny bushes	Scattered trees	
Livestock:	Ruminants Free grazing in winter	Grazing	Free grazing	Grazing		
Problems:	Pod borer in mash Water flooding from fan Water shortages	Poor soil	Serious water shortages	Poor soil	Flooding	
Opportunities:	Pest control Extend canals–irrigation and water control		Improve grazing Extend canals	More trees	Alfalfa irrigation	

Source: Gordon R. Conway, *et al.*, *Agroecosystem Analysis and Development for the Northern Areas of Pakistan* (Gilgit, Pakistan: Aga Khan Rural Support Programme, 1986).

A more recent example of RAZ was carried out in the Alpuri subdivision of Swat District in the North West Frontier Province of Pakistan.[9] It was undertaken at the beginning of a project to develop horticulture and vegetable production in the area and the general recommendation domains with their associated strategies were defined in these terms.

Village Agroecosystem Analysis
The next stage is to investigate and characterize further the RAZ zones by an AEA conducted on representative villages. The primary aim is to determine key hypotheses for research and development appropriate for each zone.

Following a secondary data review of the village the analysis begins with a field visit whose aim is to produce a series of diagrams of the agroecosystem based on direct observation and semi-structured interview. These diagrams are designed to relate to four basic system patterns – in space and time, and of flows and decision making. Maps and transects (Figure A.2) describe the spatial patterns and, in particular, the location of particular problems and opportunities. Seasonal calendars (Figure A.3) and graphs summarize patterns in time, showing interrelationships between a wide variety of activities. Sources and flows of income are summarized in bar diagrams based on semi-structured interviews. Flow diagrams are used to summarize production and marketing cycles and the actual or potential impact of major innovations or interventions (Figure A.4). Finally, decision trees describe the choices of different farming strategies and the factors affecting these choices (Figure A.5), while Venn diagrams are used to analyse institutional interactions in decision making.

The field visits take about two days per village. The diagrams are then converted to overhead transparencies and used as the focus of a half- to one-day workshop. This involves the field team and other members of the development group and is structured around the procedure shown in Figure A.6. Following a brief discussion of objectives and system boundaries the team concentrates on the analysis of the diagrams. This has two outcomes: first, as an intermediate step, a table of the most

Figure A.3: Seasonal calendar for a village in the Hunza Valley

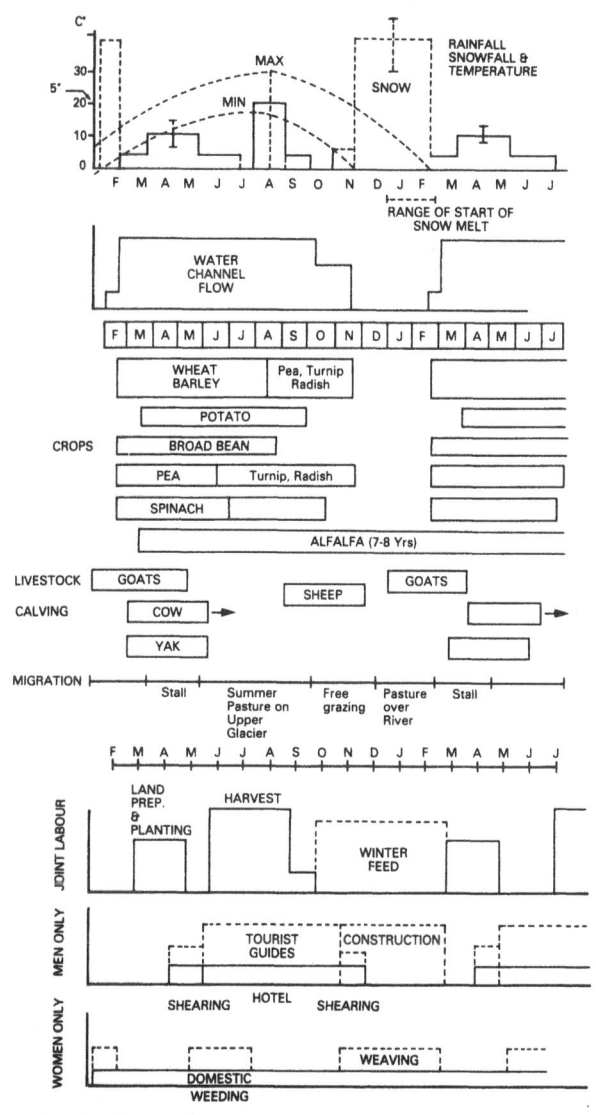

Source: Gordon R. Conway, *et al.*, *Agroecosystem Analysis and Development for the Northern Areas of Pakistan* (Gilgit, Pakistan: Aga Khan Rural Support Programme, 1986).

Figure A.4: Impact flow diagram for a village in the Hunza Valley: effect of a new highway

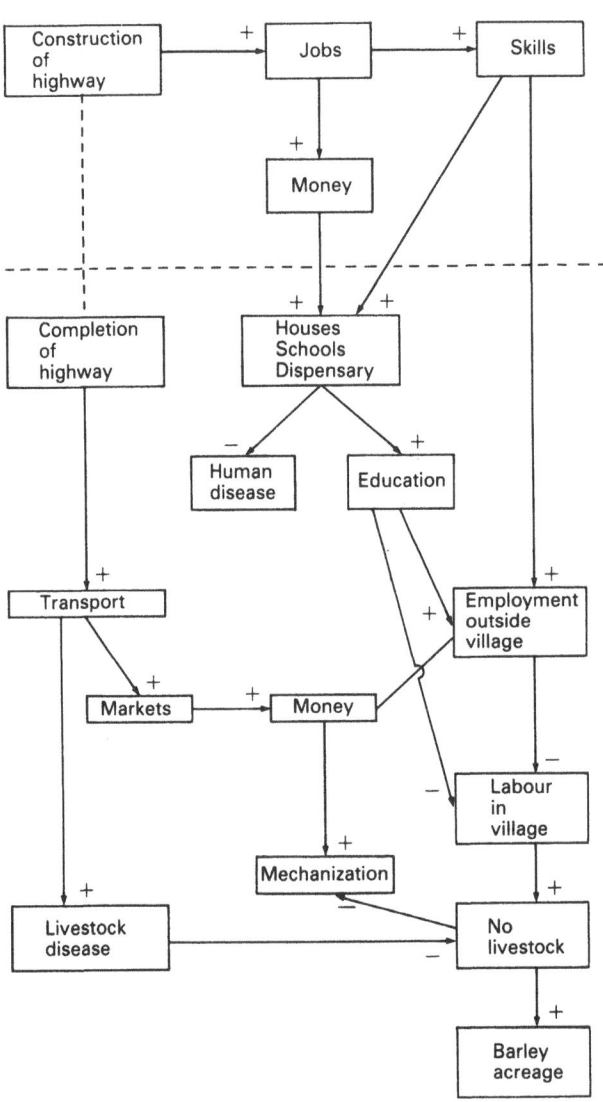

Source: Gordon R. Conway, *et al.*, *Agroecosystem Analysis and Development for the Northern Areas of Pakistan* (Gilgit, Pakistan: Aga Khan Rural Support Programme, 1986).

Figure A.5: Decision tree for livelihood systems for a village in the Hunza Valley

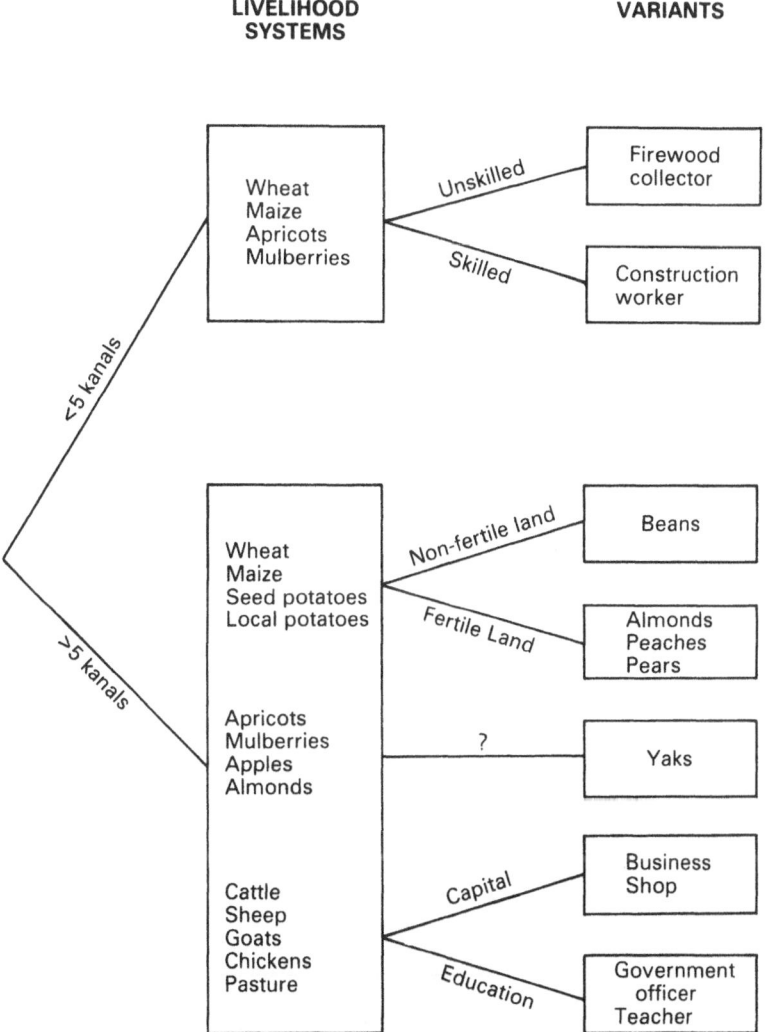

Source: Gordon R. Conway, *et al., Agroecosystem Analysis and Development for the Northern Areas of Pakistan* (Gilgit, Pakistan: Aga Khan Rural Support Programme, 1986).

Figure A.6: The procedure of agroecosystem analysis

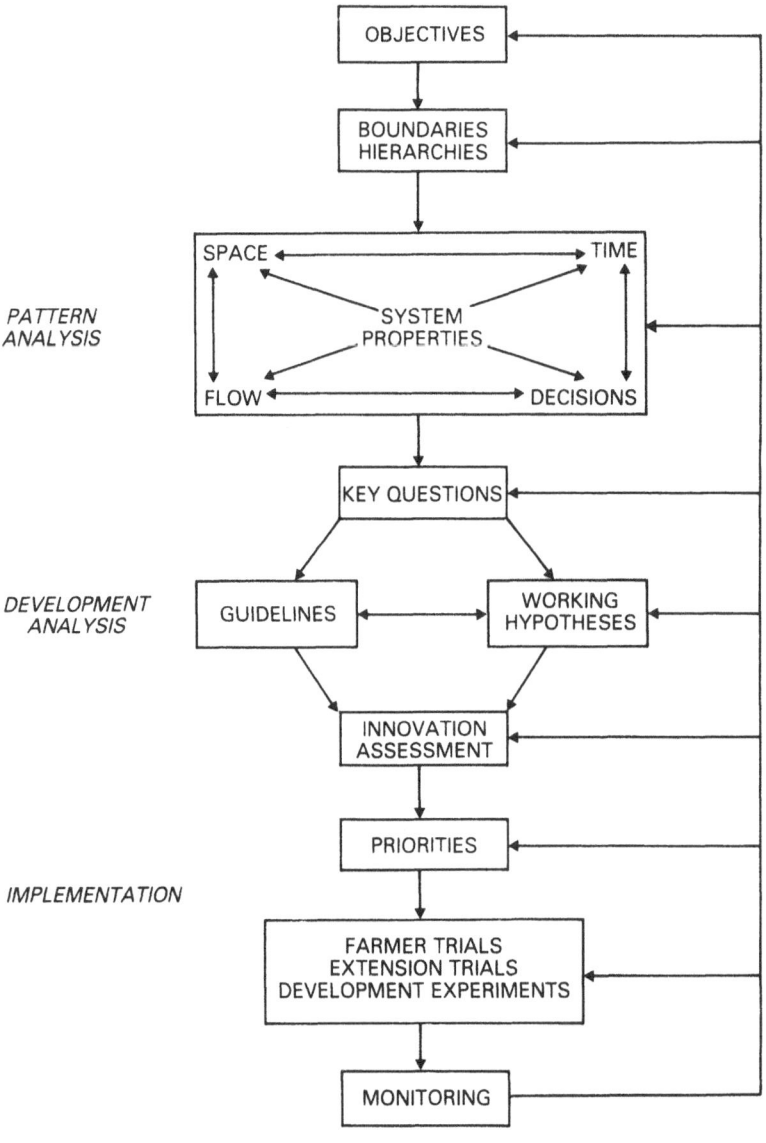

important factors affecting the system properties of productivity, stability, sustainability and equitability; and, second, a set of key questions and related hypotheses that the team as a whole agrees are of primary importance. Table A.2 provides examples of the kinds of questions that emerge.

The final stage is for the team to assess the innovations and interventions implicit in the key questions in terms of their effects on system properties, and to rank them for priority attention (Table A.3). Some of the questions are key research questions with related research hypotheses; others are key development questions with related working hypotheses. The first category leads to further research, ideally conducted with the full participation of the

---

Table A.2: Examples of key questions relating to the development of the new land in a village in northern Pakistan

Key question 1:   How can soil development be speeded up while at the same time providing a higher return on new land?

Working hypothesis: The third terrace should be planted with apples, peaches, apricots and cherries, plus alfalfa. Second terrace should be planted with willow, rubinia, alfalfa and perennial grasses

Key question 2:   How can land be used efficiently after reclamation?

Working hypothesis: After 7 years, 25% of the land should be utilized for potato and the rest planted to wheat, barley and fruit trees. Alfalfa and grasses should be planted on the second terrace

Source: Gordon R. Conway *et al., Agroecosystem Analysis and Development for the Northern Areas of Pakistan* (Gilgit, Pakistan: Aga Khan Rural Support Programme, 1986).

Table A.3: Innovation assessment for a village in the Hunza Valley

INNOVATION	Productivity	Stability	Sustainability	Equitability	Cost	Time for benefits	Feasibility	Priority
Development of second terrace	+	+	++	++	□	◪	■	3
Development of third terrace	+++	+	+	+	□	□	◪	1
Artificial insemination	++	+	?	−	◪	◪	◪	2
Catch crops	+	+	+	o	■	■	■	3
River bunds	o	+	+	+	□	■	◪	4
Potato intensification	++++	+	−	+	■	■	◪	2
Involvement of women	++	+	+	+++	■	◪	□	2

Source: Gordon R. Conway, *et al.*, *Agroecosystem Analysis and Development for the Northern Areas of Pakistan* (Gilgit, Pakistan: Aga Khan Rural Support Programme, 1986).

villagers; the second category results in development action.

Rapid Rural Appraisal

Agroecosystem Analysis can be regarded as but one example of the approach known as Rapid Rural Appraisal (RRA) which has been developed over the last decade.[10] RRA may be defined as a systematic but semi-structured activity carried out in the field by a multidisciplinary team, and designed to acquire quickly new information on, and new hypotheses about, rural life.

Two themes are central to the philosophy of RRA. The first is the pursuit of "optimal ignorance". This implies that both the amount and the detail of information required to formulate useful hypotheses in a limited period of time are regarded as expenses to be kept to a minimum. In terms of the concepts presented earlier, the aim of the multidisciplinary team is to arrive at an agreed sufficiency of knowledge of the key agroecosystem processes and properties relevant to the objectives of the RRA, and not to exceed this by investigating irrelevant aspects or being concerned with unnecessary detail.

The second theme is diversity of analysis. This is pursued through the process of "triangulation" – that is, the use of several different sources of, and means of gathering, information. Notwithstanding the self-imposed limits of time and resources, the accuracy and completeness of an RRA study is maximized by investigating each aspect of the situation in a variety of ways. "Truth" is approached through the rapid build-up of diverse information rather than via statistical replication. Secondary data, direct observation in the field, semi-structured interviews and the preparation of diagrams all contribute to a progressively more accurate analysis of the situation under investigation.

These themes in turn lead to five key features of good RRAs, namely that they are:

- *iterative* – the process and goals of the study are not immutably fixed beforehand, but modified as the team realizes what is or is not relevant
- *innovative* – there is no simple, standardized methodology.

Techniques are developed for particular situations depending on the skills and knowledge available

- *interactive* – all team members and disciplines combine together in a way that fosters serendipity and interdisciplinary insights
- *informal* – the emphasis is, in contrast to the formality of other approaches, on partly structured and informal interviews and discussions
- *in the community* – the aim is not just to gather data for later analysis. Learning takes place largely in the field "as you go", or immediately after, in short intensive workshops. In particular, farmers' perspectives are used to help define field conditions.

Studies of local rural situations in developing countries have often concentrated on only one set of conditions, investigating for instance the economic, social, ecological *or* agricultural aspects. Where several sectors are included, as in project designs, they are often still considered in isolation from each other, at best being collected together in a single voluminous report. Extensive data collections, involving many researchers over a long period of time and costing large sums of money, are often regarded as integral to the process. These are usually followed by equally extensive statistical analyses, although often remaining narrow in their focus and assumptions.

The obvious logistical problems of such an approach are frequently accompanied by other, more serious, shortcomings. Local inhabitants are seldom consulted, or at best through fixed and formal channels, for instance by means of a written questionnaire with the questions determined beforehand and unchanged from day to day of the study, or from farm to farm or village to village. The context of the target data is frequently ignored; "averages" are sought, while significant variations are often missed. This gives little opportunity for new features of the system to be revealed or for insights to be gained other than those which could have been learnt at the start from the local people.

Such inflexible methodologies are also responsible for the

collection of many irrelevant data and the disregard of local peculiarities in, for example, the ecological, economic or cultural conditions. Delay in providing the results can sometimes lead to them being useless in the "by-then-changed" situation. A general consequence is that development projects fail through a combination of incorrect knowledge and a lack of co-operation on the part of "those being developed".

The work of the early practitioners of RRA was brought together in conferences at the Institute of Development Studies, University of Sussex in October 1978 and December 1979.[11] A more recent conference was held in September 1985 at Khon Kaen University, Thailand.[12]

The suite of RRA techniques
There is no single, standardized methodology for RRA. In each situation the methodology depends on the objectives, local conditions, skills and resources. However, there is a suite of techniques in existence which can be used in various combinations to produce appropriate RRA methods. The suite includes:

- secondary data review
- direct observation
- diagrams
- semi-structured interviews
- analytical games
- portraits and stories
- workshops.

Secondary data consist of reports, maps, aerial photographs and so on, which already exist and are relevant to the project. The review process involves searching for relevant data and summarizing these in diagrammatic models, simple tables and brief abstracts. The aim is to be sceptical and critical and to look out for what has been missed, but not to spend time here that could be better spent in the field. Direct observation includes measurement and recording of objects, events and processes in the field, either because they are important in their own right or because they are surrogates for other variables which are important. Diagrams have already been described.

One of the most important of RRA techniques is semi-structured interviewing (SSI), which is a form of guided interviewing where only some of the questions are predetermined and new questions or lines of questioning arise during the interview, in response to answers from those interviewed. The information is thus derived from the interaction between the knowledge and experience of the interviewer and the interviewee(s). The latter may be groups, for example of village leaders, or key informants, such as school teachers or local government officials, or the farmers themselves, selected on one or more criteria.

Analytical games consist of dialogues with farmers which take the form of a game, i.e. they follow certain simple but mutually agreed rules. One example is "Preference Ranking"[6] where farmers are asked to choose between pairs of crop varieties or tree species. A set of choices is prepared and farmers are presented with the choices in every possible combination of two to compare. They are asked to indicate which they would choose if they could only grow one of the pair they prefer and give the reasons.

Portraits and stories are simple written essays on families and their livelihoods, which illuminate their present conditions and the manner of their decision making.

Classes of RRA

The various techniques described above will be used in various combinations depending on the objective of the RRA. Very broadly, there are four principal classes of RRA, which ideally follow one another in the sequence of development activity:

- Exploratory RRA – to obtain initial information about a new topic or agroecosystem. The output is usually a set of preliminary key questions and hypotheses. (Agroecosystem Analysis is an example of an exploratory RRA).
- Topical RRA – to investigate a specific topic, often in the form of a key question and hypothesis generated by the exploratory RRA. The output is usually a detailed and extended hypothesis that can be used as a strong basis for research or development.

- Participatory RRA – to involve villagers and local officials in decisions about further action based on the hypotheses produced by the exploratory or topical RRAs. The output is a set of farmer-managed trials or a development activity in which the villagers are closely involved.

- Monitoring RRA – to monitor progress in the trials and experiments and in the implementation of the development activity. The output is usually a revised hypothesis together with consequent changes in the trials or development intervention which will hopefully bring about improved benefits.

TOPICAL RRA

A topical RRA aims to answer specific questions on a certain topic and as such has a narrower scope of investigation than an exploratory RRA. Examples of topics investigated by this type of RRA are listed in Table A.4.

While focusing on one particular question, however, the topical RRA does *not* limit itself to only one facet of the issue. As in all RRAs, a systems approach and a multidisciplinary team are used and all the techniques described above are likely to be included. As the topical RRA proceeds, the scope of investigation narrows further, while the depth of analysis increases. Thus the general, sometimes naïve, inquiries of the literature search stage give way, through probing questions and analysis, to more considered, "optimally informed" opinions in the final stage.

Figure A.7 shows the relative duration of the techniques involved and how they are sequentially organized in a typical topical RRA. As can be seen, a possible end product of the process is an "extended hypothesis". Rather than a definitive answer to the question, this gives a concise description of the situation from the viewpoint of the researchers and summarizes the primary causes they suggest are responsible for the problem being investigated. The hypothesis can then be used in one of two ways. Firstly as a "working hypothesis" it can be assumed to be a true representation of the situation and actions can be taken, based on its findings and recommendations. Alternatively

Table A.4: Examples of topical RRAs undertaken from Khon Kaen University

Topic under investigation	*No. of participants in RRA team*
Causes and effects of trees in the paddy fields of NE Thailand	4
Fuelwood situation in NE Thailand – problems and processes of adjustment to its availability	7
Major factors explaining the various degrees of success in the operation of three dairy villages in NE Thailand	10
Cropping patterns and the use of crop residues to supplement feed in dairy calf production in NE Thailand	4
Extent of replacement of native black swine with non-native white in NE Thailand, reasons why native swine production is still practised and whether such production will be sustained in the future	12
Socio-cultural and biophysical conditions allowing farmers in Surin to adopt peanuts after rice using residual soil moisture	6
How villagers in Srisaket have adjusted to annual flooding	5
Factors responsible for the varying degrees of usage of small-scale irrigation systems in NE Thailand	4

Figure A.7: The process of a topical RRA

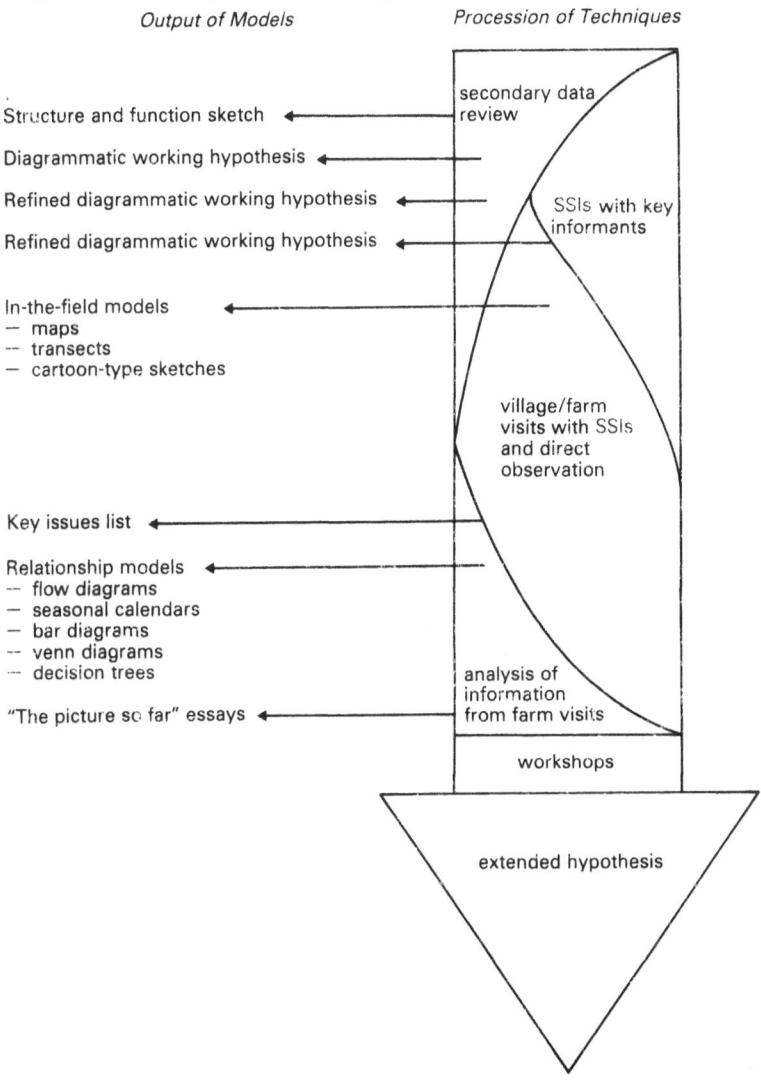

Output of Models   *Procession of Techniques*

secondary data review ← Structure and function sketch

Diagrammatic working hypothesis ←

Refined diagrammatic working hypothesis ←

Refined diagrammatic working hypothesis ←

SSIs with key informants

In-the-field models
— maps
— transects
— cartoon-type sketches

village/farm visits with SSIs and direct observation

Key issues list ←

Relationship models
— flow diagrams
— seasonal calendars
— bar diagrams
— venn diagrams
— decision trees

analysis of information from farm visits

"The picture so far" essays ←

workshops

extended hypothesis

Source: Jennifer A. McCracken,"A working framework for RRA: lessons from a Fiji experience", *Agricultural Administration*, vol.29 no.3 (1988), pp.163-84.

it can be considered a "research hypothesis" and research set up to confirm or disprove it; this in turn may lead to recommendations for action.

The way in which ideas are "filtered and focused" to arrive at the hypothesis (i.e. the thought processes behind the framework of Figure A.7) will vary according to the topic of the RRA. Up to now no attempt has been made to produce a standardized format for this aspect of all topical RRAs. However, an RRA exercise carried out in Fiji attempts to bridge this gap and provide future practitioners with a practical structure on which to base their own individual initiatives.[13] This RRA aimed to answer the key question: "Why are sugarcane yields in Fiji low?". Firstly the secondary data survey, including analysis of international cane yield reports and farm data files of the Fiji Sugar Corporation, addressed the preliminary, broadbased questions of: "Is there a problem of low cane yields in Fiji?" and "Are the yields uniformly low over Fiji?". Having found the answers to these questions to be "Yes" and "No" respectively, the study went on to try and identify the causal factors of the low yields and their spatial variation.

A range of the possible causal factors and processes were laid out as diagrammatic hypotheses (Figure A.8) which were then refined and modified through SSIs with key informants and cane growers, direct observation in the field and the production of diagrammatic models. From the final version of the diagrammatic hypothesis, a short list of key causal factors was distilled and this then became the base for deeper analyses. All the information, ranging from the secondary data to the opinions expressed in the interviews, was brought together for each factor and a series of brief essays produced. The titles of these essays included: "Cane farming in Fiji", "Patterns of farming in Lovu" (the survey area), "Differences between Lovu growers", "Climate", "Poor management", and "Poor soil and slope". Finally the filtering process culminated in a single extended hypothesis which brought together all the causal factors and put forward a concise description of the problem situation of Fiji cane yields.

Its basic thesis was:

The cane yield of each farm depends on the balance between farm

Figure A.8: Diagrammatic working hypothesis for influence of farming practices on low cane yields in Fiji

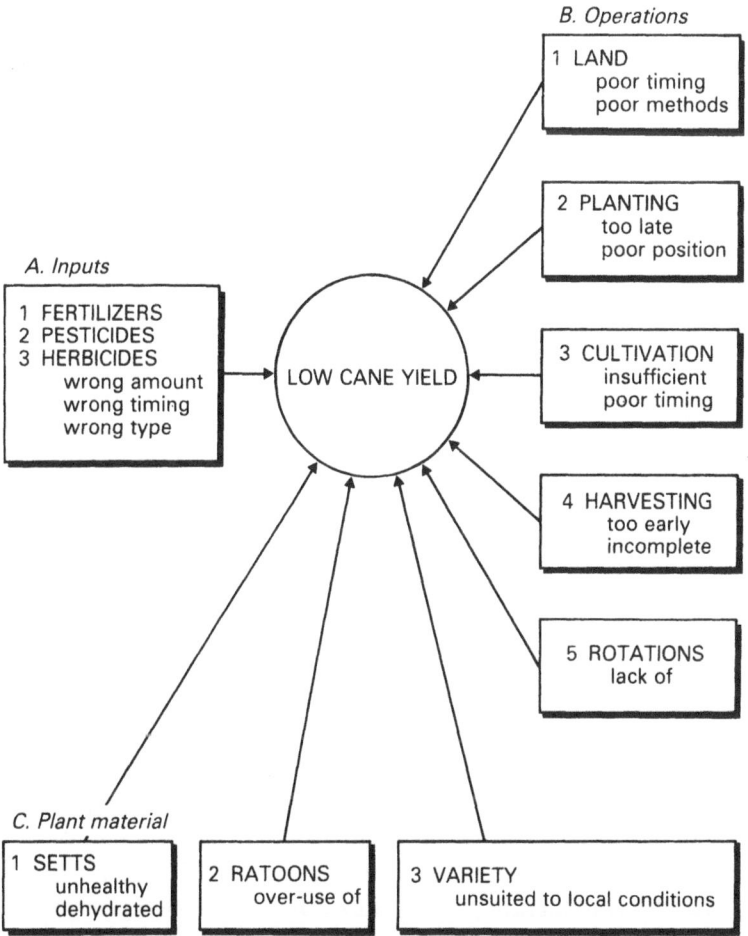

Source: Jennifer A. McCracken, "A working framework for RRA: lessons from a Fiji experience", *Agricultural Administration*, vol.29 no.3 (1988), pp.163-84.

size and household size. Since the vast majority of farms are too small to support the average household by their cane production alone, at least some of the members must engage in off-farm work. The inevitable neglect of the cane means the majority of Fiji's cane producers obtain low yields. Medium-sized farms allow room for the production of vegetables, rice and goats in addition to the cane. This has two effects – the marketing of the produce lessens the household's cash flow problems and the need for much more regular farmwork keeps more family members on the farm and available for cane work when required. Such farms' cane yields are generally higher than the national average. The largest farms, while benefiting from the opportunity for diversified production generally have cane areas too large to be managed by the household alone and thus require the expense of labour hire. Location of the farms is also an important factor – these largest farms are on the hilliest, remotest and least fertile land and so the growers incur extra costs of machinery hire for land levelling, transport hire for taking the cane to the mill, and large amounts of fertiliser. These large farms rarely produce high yields of cane.[13]

PARTICIPATORY RRA

All RRA exercises have at least some element of participation by the farmers and rural poor on whom they are targeted. At the very least they are partners in semi-structured interviewing. However, there will be many situations in which the primary goal is to involve the local people in crucial decision-making. Then specifically designed participatory RRAs may be appropriate.

An example of a participatory RRA is the approach adopted by the Aga Khan Rural Support Programme (AKRSP), and referred to earlier. The AKRSP is concerned with developing several hundred villages in the northern areas of Pakistan in ways which capitalize on local knowledge and skills, and accord with local wishes.[14] The first phase of the programme, begun in 1983, focused on the development in each village of a single physical infrastructure project. The classical project cycle is here accomplished through a series of interactive dialogues, termed the diagnostic survey. The first dialogue is conducted by the project management team in the village with an assembly of villagers. They are asked to identify one infrastructure project which will increase village incomes, and which can be implemented and maintained by them. One condition is that the project must benefit at least 70–80% of the village population. Once a project is identified (perhaps after much debate and many assemblies), the second dialogue occurs. This involves project engineers who survey the site, draw up plans and estimate costs, all with the participation of knowledgeable villagers nominated by the village at the first dialogue. After the plans and estimates are checked by the senior engineer, project management returns to the village for a third dialogue. This consists of a full discussion of the rights and responsibilities of both AKRSP and the villagers. Terms of partnership are drawn up which explain the basis, amount and method of payment, and the common responsibilities of the village in managing the project and its finances, manpower and material. The project is initiated only if villagers can demonstrate their capacity for managing it on a permanent basis. A good part of the project cost is funded by a one-time grant by AKRSP to the village as a whole.

The diagnostic survey of the entire project area in Gilgit District (more than 300 villages) resulted in the initiation of 256 village-level infrastructure projects (generally one per village) at a cost of Rs38.0 million (US$2 million). So far 180 projects have been completed and an estimated 22,098 families will benefit. Only ten projects are considered to be moving "slowly", and none have fallen into disuse through lack of repairs.

The diagnostic survey and the ensuing funding of the physical infrastructure projects provided an entry point for the AKRSP to assist with development of the villages. This has opened the way for the second phase of development which focuses on realizing the agricultural potential of the villages through the judicious use of loans made against the villagers' savings. It is at this stage that Agroecosystem Zoning and AEA have been used.

MONITORING RRA
The aim of a monitoring RRA is to assess the impact of an intervention on a particular agroecosystem. Such intervention may have been the result of previous findings and recommendations from exploratory, topical or participatory RRA.

All the RRA techniques can be usefully applied in monitoring RRA. Secondary data review, particularly of the initial project documentation, is crucial to producing clear "before and after" comparisons. The central feature of a monitoring RRA, however, is visits to the target area, which involve direct observation of the changes that have occurred, and semi-structured interviews with the local inhabitants. Besides assessment of direct productivity indicators, the interviews should provide information on changes in life-styles, livelihoods, level of independence, contact with other areas, and also on the opinions of those affected.

Where the intervention being monitored has run into problems, and particularly where conflicts of interest have emerged, a workshop may be useful. This can bring together representatives of those affected in the different areas and at different levels, along with the planners and implementers of the intervention. Experiences can be compared and any knock-on effects traced. The effects, social and economic, short-term and long-term, local and widespread, can be viewed together. Matrices can be drawn up to identify the relationships between these different components of the system and the positive and negative consequences for each. Following these discussions, recommendations can be agreed on how the intervention could be improved or any adverse effects alleviated.

An example of such a workshop was held in the Philippines, concerning the effects of a major development project – The Bicol Integrated Area Development III (Rinconada/Buhi-Lalo) Project.[15] This project, as part of a larger scale programme of river basin development, had already been found to have adverse effects on the fishing and agricultural livelihoods of the lakeshore inhabitants, the transportation and domestic water supplies of a lakeside town, and the long-term productivity and stability of the lake itself. The workshop was a modification of the Agroecosystem Analysis approach. Over a period of five days, government officials, development agency personnel and representatives of the farmers and fishermen came together to analyse a series of summary diagrams prepared during a brief field visit by a project team. The outcome was a set of key questions and hypotheses for research and development which

Figure A.9: Models for project design and implementation

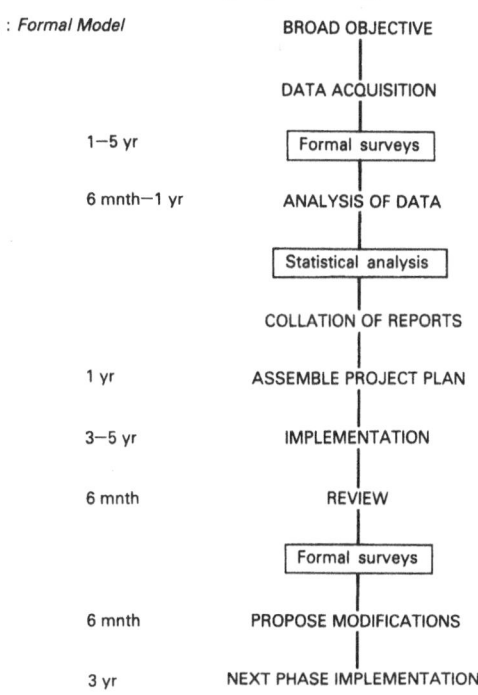

(a) : *Formal Model*

	BROAD OBJECTIVE
	DATA ACQUISITION
1–5 yr	Formal surveys
6 mnth–1 yr	ANALYSIS OF DATA
	Statistical analysis
	COLLATION OF REPORTS
1 yr	ASSEMBLE PROJECT PLAN
3–5 yr	IMPLEMENTATION
6 mnth	REVIEW
	Formal surveys
6 mnth	PROPOSE MODIFICATIONS
3 yr	NEXT PHASE IMPLEMENTATION

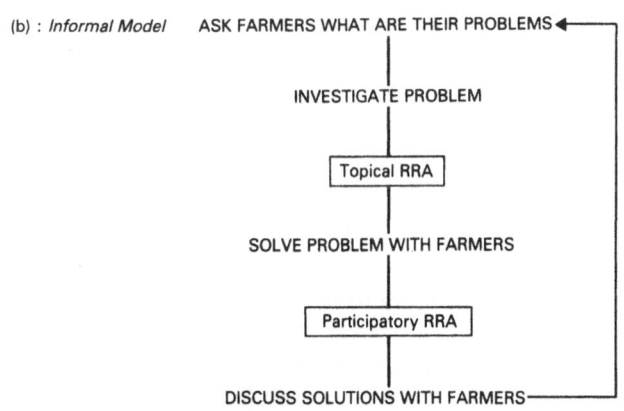

(b) : *Informal Model*

ASK FARMERS WHAT ARE THEIR PROBLEMS

INVESTIGATE PROBLEM

Topical RRA

SOLVE PROBLEM WITH FARMERS

Participatory RRA

DISCUSS SOLUTIONS WITH FARMERS

were then assessed by the whole workshop to produce a plan of recommended changes and improvements.

A monitoring RRA, including such a workshop, need not take any longer than other forms of RRAs. Several such short surveys can be conducted over a period of time to monitor the progress of a project and to make corrections and improvements.

Project design and implementation

A traditional project cycle, particularly as it applies to large investment projects, goes through the sequential phases of data acquisition, analysis, planning, implementation, review and redesign (Figure A.9(a)). It is ordered and methodical, but is often a costly and time-consuming exercise. The logical progression is one which is designed to ensure that all factors and considerations are incorporated. But, as experience has shown, such an approach tends to become rigid and bureaucratized in practice. Critical questions are not asked and important insights are missed. At the other extreme is a project cycle which simply moves rapidly from identifying problems to solving them, and back again (Figure A.9(b)). This approach is based on a very close relationship between development professionals and farmers and can only really be undertaken on a small and intimate scale.

This latter scheme relies entirely on RRAs. However, RRA has its limitations. It will never, and indeed was never designed to, make redundant more traditional, formal and detailed surveys and analyses. RRAs and RRA techniques essentially complement more formal methods and while in some situations they may be substitutes, more often than not, they are preliminary exercises, leading up to more detailed analyses.

The advent of RRA has thus greatly enriched the availability of methods of analysis for rural development. Techniques can be chosen on the basis of the nature of the problem, the local situation and the resources to hand. In particular, different techniques, both formal and informal, can be blended to produce a project cycle along the lines of Figure A.10. This lies some way between the extremes of the schemes in Figure A.9, and can be applied to a wide range of projects, both large and small.

In such a scheme the primary role of the RRA is to define
and refine hypotheses which are then tested, either formally or
informally, as part of the project cycle. Providing the cycle is
iterative, flexible and open, it should be possible to combine
speed with both rigour and sensitivity resulting in development
that is not only productive but durable and equitable in its benefits.

**Figure A.10: A model for project design and implementation
which combines the use of Rapid Rural Appraisal and formal
analysis and survey**

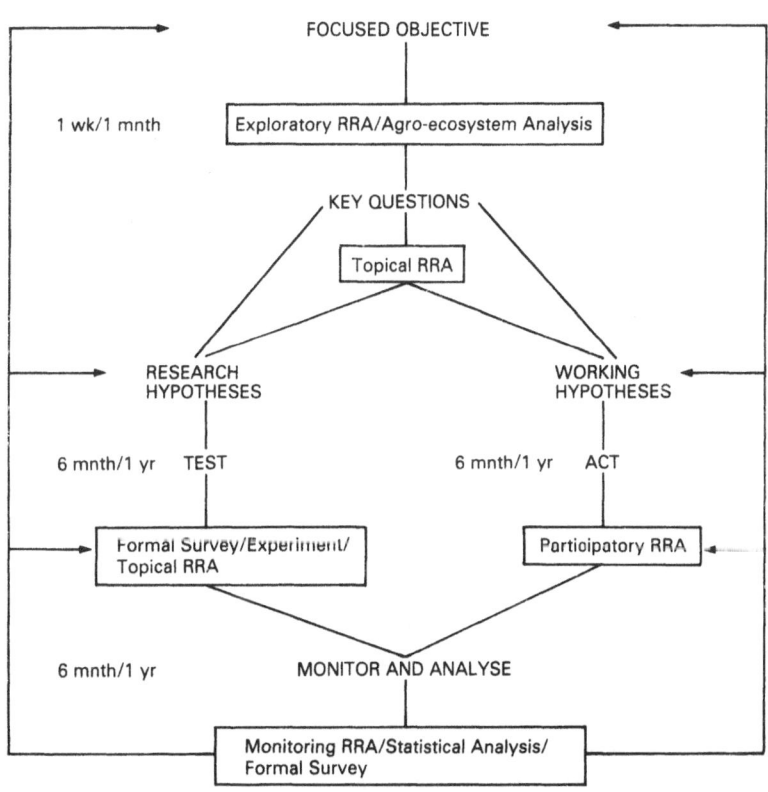

Notes

1. See Gordon R. Conway, "Agroecosystem analysis", *Agricultural Administration*, vol.20 (1985), p.31; Gordon R. Conway, *Agroecosystem Analysis for Research and Development* (Bangkok: Winrock International Institute for Agricultural Development, 1986).

2. Phrek Gypmantasiri *et al.*, *An Interdisciplinary Perspective of Cropping Systems in the Chiang Mai Valley: Key questions for research* (University of Chiang Mai, Thailand: Faculty of Agriculture, 1980).

3. KKU–Ford Cropping Systems Project, *An Agroecosystem Analysis of Northeast Thailand* (Khon Kaen: Faculty of Agriculture, University of Khon Kaen, 1982); KKU–Ford Cropping Systems Project, *Tambon and Village Agricultural Systems in Northeast Thailand* (Khon Kaen: Faculty of Agriculture, University of Khon Kaen, 1982); V. Limpinuntana and A. Patanothai, *Handbook of the NERAD Tambons* (Khon Kaen: Northeast Rainfed Agricultural Development Project, Northeast Regional Office of Agriculture and Co-operatives, Tha Phra, 1984).

4. KEPAS, *The Critical Uplands of Eastern Java: An agroecosystem analysis* (Jakarta: Kelompok Penelitian Agro-Ekosistem, Agency for Agricultural Research and Development, 1985); KEPAS, *Swampland Agroecosystems of Southern Kalimantan* (Jakarta: Kelompok Penelitian Agro-Ekosistem, Agency for Agricultural Research and Development, 1985); KEPAS, *Agro-ekosistem Daerah kering di Nusa Tenggara Timur* (Jakarta: Kelompok Penelitian Agro-Ekosistem, Agency for Agricultural Research and Development, 1986).

5. Gordon R. Conway *et. al.*, *Agroecosystem Analysis and Development for the Northern Areas of Pakistan* (Gilgit, Pakistan: Aga Khan Rural Support Programme, 1986); Gordon R. Conway *et. al.*, "Rapid Rural Appraisal for sustainable development: experiences from the northern areas of Pakistan", in Czech Conroy and Miles Litvinoff (eds), *The Greening of Aid* (London: Earthscan Publications, 1988).

6. Ethiopian Red Cross Society, *Rapid Rural Appraisal: A closer look at rural life in Wollo* (Addis Ababa: Ethiopian Red Cross Society) and (London: IIED).

7. Klingbiel and Montgomery, *Land Capability Classification* (Washington, DC: USDA Soil Conservation Service); G. Higgins and Amir H. Kassam, "The FAO agro-ecological approach to determination

of land potential", *Pedalogie*, vol.31 (1981), pp.147–48; and C.R. Holdridge, *Life Zone Ecology* (Costa Rica: Tropical Science Centre).

8. Gordon R. Conway *et. al.*, in Conroy and Litvinoff (eds), op. cit.

9. Malakand Fruit and Vegetable Development Project, *Rapid Agro-ecosystem Zoning of Alpuri, Pakistan*, J.N. Pretty (ed.), (MFVDP, PO Box 21, Saidu Sharif, Swat, Pakistan, 1988; and IIED, London).

10. See Jennifer A. McCracken, Jules N. Pretty and Gordon R. Conway, *An Introduction to Rapid Rural Appraisal for Agricultural Development* (London: IIED, 1988), for a general introduction and an annotated bibliography; also Gordon R. Conway and Jennifer A. McCracken, "Rapid Rural Appraisal and Agroecosystem Analysis", in M.A. Altieri and S.B. Hecht (eds), *Agroecology and Small Farm Development*, (Florida: CRC Press, 1990).

11. See papers in *Agricultural Administration*, vol.10 (1981).

12. Khon Kaen University, *Proceedings of the International Conference on Rapid Rural Appraisal*, Khon Kaen, Thailand, 2–5 September 1985 (1987).

13. Jennifer A. McCracken, "A working framework for RRA: lessons from a Fiji experience", *Agricultural Administration*, vol.29, no.3, pp.163–84.

14. Tariq Husain, *Approaches to Understanding Mountain Environments, Research and Planning Experiences from the Gilgit District of Pakistan* (Gilgit, Pakistan: Aga Khan Rural Support Programme, 1987).

15. Gordon R. Conway and Percy J. Sajise (eds), *The Agroecosystems of Buhi: Problems and opportunities*, (Los Banos: University of the Philippines, 1986); and Gordon R. Conway, Percy J. Sajise and William Knowland, "Lake Buhi: resolving conflicts in a Philippine development project", *Ambio*, vol.18, (1989), pp.128–35.

Index